Homesteading in Your Backyard
Harnessing Nature's Bounty Right Outside Your Door

Michael Thompson

© Copyright 2024 - All rights reserved.

The content contained within this book may not be reproduced, duplicated or transmitted without direct written permission from the author or the publisher.

Under no circumstances will any blame or legal responsibility be held against the publisher, or author, for any damages, reparation, or monetary loss due to the information contained within this book, either directly or indirectly.

Legal Notice:

This book is copyright protected. It is only for personal use. You cannot amend, distribute, sell, use, quote or paraphrase any part, or the content within this book, without the consent of the author or publisher.

Disclaimer Notice:

Please note the information contained within this document is for educational and entertainment purposes only. All effort has been executed to present accurate, up to date, reliable, complete information. No warranties of any kind are declared or implied. Readers acknowledge that the author is not engaging in the rendering of legal, financial, medical or professional advice. The content within this book has been derived from various sources. Please consult a licensed professional before attempting any techniques outlined in this book.

By reading this document, the reader agrees that under no circumstances is the author responsible for any losses, direct or indirect, that are incurred as a result of the use of information contained within this document, including, but not limited to, errors, omissions, or inaccuracies.

Table of Contents

INTRODUCTION 5

CHAPTER I. Getting Started with Backyard Homesteading 7

    Definition and philosophy of homesteading 7

    Benefits of backyard homesteading 12

    Assessing your space and resources 17

    Planning and goal setting 22

CHAPTER II. Soil Preparation and Gardening Basics 31

    Techniques for improving soil fertility 31

    Composting and vermiculture 41

    Planning your garden layout 47

CHAPTER III. Growing Your Own Vegetables and Herbs .. 53

    Choosing the right plants for your climate 53

    Starting from seeds vs. transplants 60

    Seasonal planting guide 65

    Tips for maximizing yield in small spaces 70

CHAPTER IV. Fruits, Berries, and Nut Trees 77

    Selecting fruit trees and bushes for your backyard 77

    Planting and caring for perennial plants 83

    Harvesting and preserving fruits and nuts 89

CHAPTER V. Backyard Livestock 96

    Benefits and considerations for raising animals............ 96

    Chickens: breeds, care, and egg production 103

Bees: starting a beehive and honey production 110
CONCLUSION **115**

INTRODUCTION

Through useful homesteading techniques, "Homesteading in Your Backyard: Harnessing Nature's Bounty Right Outside Your Door" introduces readers to the rewarding realm of self-sufficiency and sustainable living. This all-inclusive manual is designed for anyone, regardless of space limitations or expertise level, who wants to turn their backyard into a profitable and satisfying farm. Through utilizing the nearby natural resources, readers set out on a path to enhanced resilience, ecological responsibility, and a more profound bond with the land.

The first section of the book covers the fundamental ideas of homesteading, stressing the value of independence, sustainability, and an all-encompassing lifestyle. It inspires readers to live more simply while making the most of their outside areas, whether it be through backyard beehives for honey production, gardening, or rearing small livestock like chickens or rabbits. Every chapter equips readers to launch and manage their own homestead projects with a wealth of useful tips, detailed instructions, and original ideas.

In "Homesteading in Your Backyard," readers will learn creative ways to cultivate organic fruits, vegetables, and herbs in small spaces by utilizing vertical and container gardening. It explores the skill of composting, the health of the soil, and organic pest control methods that support a flourishing garden ecology. The book offers insights for anybody interested in animal husbandry, such as managing small livestock such as goats or rabbits for milk, meat, or fiber, growing birds for eggs and meat, as well as maintaining effective beehives for honey production.

In addition to emphasizing useful skills, "Homesteading in Your Backyard" highlights the holistic advantages of homesteading, such as building a feeling of community and enhancing mental and physical health. It inspires readers to adopt a more sustainable lifestyle that is consistent with their goals and values, get back in touch with the natural world, and lessen their impact on the environment.

"Homesteading in Your Backyard" offers readers who live in suburbs looking for increased self-sufficiency or in cities with little outside space a complete resource and motivational partner on the path to a more resilient, satisfying, and sustainable existence.

CHAPTER I

Getting Started with Backyard Homesteading

Definition and philosophy of homesteading

Fundamentally, homesteading is a self-sufficient way of living that includes subsistence farming, food storage at home, and the basic creation of clothing, textiles, and crafts. Its origins can be traced to the Homestead Act of 1862, which encouraged American westward migration by providing land to residents who were prepared to settle on it and farm it. But this historical background has changed dramatically, and contemporary homesteading is about adopting a lifestyle that prioritizes sustainability, independence, and a stronger relationship with the natural world rather than about colonizing new territory.

Fundamentally, homesteading is characterized by the deliberate decision to lead a more straightforward and independent existence. It includes cultivating one's own food through harvest preservation, livestock keeping, and gardening. It also includes activities like using renewable resources, making handmade things, and generating one's own energy. The goal of homesteading is to thrive by resource conservation and self-reliance, not only to survive. It encourages a way of living that aims to become more independent of commercial systems and less dependent on them.

The homesteading ideology is firmly based on a number of fundamental ideas, including self-reliance, sustainability, simplicity, and connection to the earth. The homesteader's way of life is governed by these values, which have an impact on both short- and long-term objectives. Among these, sustainability is arguably the most important, stressing the importance of living in balance with the environment. The goal of homesteading is to manage resources such that future generations can continue to use them. This includes using renewable energy sources, conserving water, composting, and engaging in organic gardening. Homesteaders seek to reduce their ecological imprint and improve the condition of the world by emphasizing sustainability.

Another tenet of the homesteading mindset is simplicity. Living with less and concentrating on what is genuinely important and required is encouraged by this idea. It is an opposition to materialism and the never-ending quest for financial gain. Homesteaders, on the other hand, aim to develop a life full of experiences and abilities. They place high importance on the caliber of their food, the fulfillment that comes from making their own products, and the peace of mind that comes from being more in one with nature. In homesteading, simplicity is about making the most of what one has and spending time and energy on things that improve self-sufficiency and well-being.

A key component of homesteading is self-reliance, which emphasizes how important it is to be able to support oneself and one's family. This idea entails learning a wide range of abilities, from basic carpentry and food preservation to gardening and animal husbandry. Homesteaders frequently pick up skills like tool maintenance, clothing making, and building essential structures on their land. Homesteaders are better able to overcome obstacles because of their self-reliance, which also builds resilience and a sense of empowerment. Additionally, it lessens reliance on outside markets and institutions, which is advantageous when resources are scarce or the economy is unstable.

A key component of homesteading is having a connection to the land. This idea acknowledges the inherent worth of the natural world as well as the significance of properly managing the land. Homesteaders form a close bond with their surroundings by learning about the seasonal patterns, the requirements of the soil, and the habits of the local species. This link encourages behaviors that support biodiversity and ecological health by fostering a sense of duty and respect for the natural world. Regenerative agriculture practices, which improve soil fertility, stop erosion, and support a healthy ecosystem, are frequently given top priority by homesteaders.

The dynamic and flexible practice of modern homesteading can be adapted to a range of lifestyles and living conditions. Modern homesteading can occur in urban, suburban, or rural environments despite the fact that traditional homesteading may conjure images of large rural farms and plenty of land. To cultivate their own food, urban homesteaders, for example, may use tiny backyard areas, communal plots, and rooftop gardens. They might also do things like raise chickens, keep bees, or use solar power to create electricity. Suburban homesteaders can reduce waste by creating composting settings, installing rainwater harvesting systems, and turning lawns into productive gardens.

Homesteading is also distinguished by its focus on the community. Although independence is a fundamental value, homesteaders frequently establish robust networks with other like-minded people. These communities pool their resources, expertise, and knowledge to form a support network that raises each member's level of overall resilience. Homesteading circles frequently hold workshops, seed exchanges, and cooperative projects that promote a spirit of cooperation and support among members. This sense of community is essential because it enables homesteaders to share knowledge and work together to overcome obstacles.

The homesteading concept places a strong emphasis on rejecting mass production and industrialized agriculture. Organic and permaculture farming practices are frequently preferred by homesteaders over traditional farming practices, which mostly rely on artificial chemicals and monocultures. Homesteaders are especially fond of permaculture, an ecologically based design system. It places a strong emphasis on developing self-sustaining ecosystems that imitate natural processes in order to decrease the need for outside inputs and increase biodiversity. This method yields better, more nutrient-dense food while simultaneously helping the environment.

Becoming a homesteader also means committing to moral production and consumption. Beyond just food, this idea also applies to clothing, home goods, and energy sources. Purchasing used goods, fixing rather than replacing broken items, and selecting eco-friendly products are among the top priorities for homesteaders. This moral course of action is consistent with the more general objective of lessening one's ecological footprint and advancing sustainability. A lot of homesteaders also work on crafts and do-it-yourself projects, producing handmade goods free from the drawbacks of mass production.

Essentially, homesteading involves developing a comprehensive way of living that combines useful knowledge with an awareness-based mindset. Although there is a large time, effort, and learning curve involved, there are numerous benefits. Since they can support themselves and live according to their principles, homesteaders frequently express a deep sense of satisfaction. In addition to promoting a more deliberate and aware way of living, the lifestyle cultivates a profound appreciation for the natural world and life's cycles.

Homesteading can also be a way to rebel against the unsustainable habits of the dominant consumption society. Homesteaders choose to make their own products, cultivate their own food, and generate their own energy instead of relying on systems that frequently put profit before sustainability and well-being. Even a small amount of opting out makes a significant statement about the value of quality over quantity and living in harmony with the environment.

Although homesteading has its challenges, it is flexible and may be scaled to fit an individual's needs and circumstances. As they acquire confidence and experience, beginners can progressively expand their efforts, starting with a small garden or a few poultry. The secret is to approach homesteading with an open mind, a desire to pick up new skills, and a dedication to lifelong learning. Although mistakes and failures are unavoidable, they also present homesteaders with priceless teaching moments that help them develop their resilience and grow as people.

In summary, self-sufficiency, sustainability, simplicity, and a strong bond with the land are at the center of the definition and philosophy of homesteading. It's a way of living that empowers people to take charge of their own health, lessen their environmental impact, and coexist peacefully with the natural world. A meaningful and rewarding way of life that is rich in community, skills, and personal fulfillment is provided by homesteading. It questions prevailing consumer culture standards and offers a workable route to a future that is more robust and sustainable. Homesteading, whether done on a little urban plot or a large rural farm, enables people to make use of the abundance of nature directly outside their door and build a self-sufficient and incredibly connected life.

Benefits of backyard homesteading

Within the boundaries of one's own property, backyard homesteading is a burgeoning movement that promotes sustainability, self-sufficiency, and a return to traditional living methods. Numerous advantages come with this lifestyle, including reduced costs, better health, more community ties, and environmental stewardship. By being aware of these advantages, more people will be encouraged to adopt backyard homesteading and realize the full potential of their living areas.

The financial savings that backyard homesteading provides are among its most evident and immediate advantages. Farmers who cultivate their own fruits, vegetables, and herbs can save a lot of money on groceries. The money saved by harvesting produce grown in-house frequently rapidly outweighs the initial cost of seeds, plants, and gardening tools. Further reducing food expenditures is the growing of small livestock, such as rabbits for meat or hens for eggs. The savings can add up over time, especially for families on a tight food budget.

Backyard homesteading has financial benefits in addition to revenue potential. You can sell extra produce, eggs, honey, and handmade goods at neighborhood farmers' markets, community-supported agriculture (CSA) programs, or to neighbors. This supports small-scale, sustainable agriculture, which boosts the local economy in addition to bringing in more cash. To further increase their income potential, some homesteaders even grow their businesses to include value-added goods like candles, pickles, jams, and soaps.

Benefits to health are yet another important aspect of backyard homesteading. A diet high in vitamins, minerals, and antioxidants is ensured by growing and eating fresh, organic vegetables. This can enhance general health and lower the risk of chronic diseases. Because they have more control over the application of fertilizers and pesticides, homesteaders frequently choose natural and organic techniques, which produce safer and cleaner food. Taking care of livestock, gardening, and farm tasks all require physical exercise that improves cardiovascular health, muscle strength, and weight control in addition to physical fitness.

Benefits to mental health are similarly significant. It has been demonstrated that gardening lowers stress, anxiety, and depressive symptoms. Even in a backyard, spending time in nature fosters tranquility and well-being. A sense of purpose and success is fostered by taking care of animals, planting seeds, and seeing homestead projects through to completion. Furthermore, the deliberate, slower pace of homesteading life can serve as a pleasant diversion from the fast-paced, technologically-driven modern world, reducing symptoms of burnout and overstimulation.

At the core of the backyard homesteading movement are environmental benefits. Homesteaders lessen their carbon footprint by cultivating their own food because a large amount of greenhouse gas emissions come from moving food from fields to stores. Reducing environmental effects is further achieved by using organic and sustainable agricultural techniques, such as crop rotation, composting, and permaculture. Water conservation is a top priority for homesteaders, who preserve this essential resource by using effective irrigation methods and rainwater-gathering equipment.

An additional environmental advantage of backyard homesteading is biodiversity. A vast range of plants, including native and heirloom varieties, are usually grown by homesteaders, supporting regional ecosystems and encouraging genetic diversity. In contrast, monocultures used in industrial agriculture have the potential to deplete soil nutrients and increase crop susceptibility to pests and diseases. Furthermore, pollinators like bees and butterflies need habitats to remain healthy ecosystems and produce food, which is why backyard homesteaders frequently establish these areas.

Additionally, backyard homesteading strengthens one's bond with the natural world. Homesteaders gain a greater understanding of the environment and its intricacies by participating in and observing nature's cycles. This link promotes more conscientious and environmentally friendly lifestyle choices, like recycling, cutting back on trash, and using less energy. A lot of homesteaders use renewable energy sources, such as wind turbines and solar panels, to lessen their environmental impact and dependency on fossil fuels.

Building communities is yet another advantage of backyard homesteading. Like-minded people frequently establish networks among homesteaders, where they exchange information, materials, and support. These linkages may contribute to the growth of robust, resilient communities that are better able to handle difficulties and emergencies. Cooperative enterprises, seed exchanges, and community gardens foster social cohesiveness and give individuals of all ages the chance to grow and learn from one another. In urban and suburban settings, where neighbors might usually stay apart, ths sense of community can be especially beneficial.

There are lots of educational opportunities with backyard homesteading. Important life skills, including food preservation, gardening, animal husbandry, and the creation of renewable energy, can be taught to both adults and children. These abilities promote a sense of empowerment and competence in addition to improving self-sufficiency. By sharing their expertise via seminars, workshops, and online forums, homesteaders frequently advance the cause of sustainable living. This information sharing contributes to the preservation of traditional knowledge and abilities that could otherwise disappear in the current era.

Moreover, backyard homesteading may contribute to food security. Homesteaders are less susceptible to price swings and supply chain interruptions because they grow their own food. This resilience is especially crucial during periods of natural disasters or unstable economies. Furthermore, homesteading techniques can aid in addressing urban areas' "food deserts," which are places with poor access to wholesome, fresh food. Urban homesteads and community gardens can supply low-income neighborhoods with locally grown, reasonably priced produce, enhancing their general health and nutrition.

It's important to recognize the aesthetic and recreational advantages of backyard homesteading. A well-kept garden and a well-planned homestead can add to the attraction and beauty of a person's land. Working on farm tasks and gardening can be rewarding pastimes that promote contentment and relaxation. Whether one is planting seeds, gathering veggies, or just taking in the results of one's effort, spending time outside provides a respite from technology and a chance to re-establish a connection with nature.

A sustainable and minimalist lifestyle is also promoted by homesteading. Rather than buying new things, homesteaders frequently embrace a frugal mindset by looking for methods to reuse and recycle materials. This encourages a circular economy and lessens waste. For instance, leftover wood can be made into furniture or fences, old pallets can be used to create garden beds, and glass jars can be used again to store food. This resourcefulness reduces the negative effects of consumption on the environment in addition to saving money.

In addition, homesteading in the backyard helps strengthen family and individual resilience. A self-sufficient lifestyle and the acquisition of a wide range of practical skills make homesteaders more equipped to deal with unforeseen circumstances and emergencies. In the event of a power outage, financial hardship, or interruption in the supply chain, homesteaders possess the necessary knowledge and resources to provide for their families and themselves. Resilience like this promotes mental stability and security.

Preserving culture is an additional benefit of backyard homesteading. Many of the customs associated with homesteading have been handed down through the ages. Homesteaders contribute to the preservation of cultural heritage and the survival of age-old skills by participating in these activities. Maintaining a sense of identity and a link to the past, as well as teaching the next generation about sustainable and self-sufficient living, depends on this cultural continuity.

Aside from all of these advantages, backyard homesteading supports larger social objectives like environmental preservation and sustainability. As more people take up homesteading, there may be a big effect on biodiversity promotion, resource conservation, and greenhouse gas emission reduction all at once. This grassroots movement backs international initiatives aimed at halting climate change and safeguarding the environment for the next generations.

To sum up, there are a wide range of advantages to backyard homesteading, including financial savings, better health, environmental stewardship, community development, education, food security, and individual resilience. People can build a more self-sufficient, sustainable, and satisfying way of life by adopting this lifestyle. In addition to improving people's quality of life, backyard homesteading benefits society as a whole by encouraging sustainable lifestyle choices and building better, more resilient communities. Homesteading offers a way of life that is richer, more purposeful, and respects both tradition and innovation, whether it is on a little urban plot or a huge suburban yard.

Assessing your space and resources

A crucial first step in starting a backyard homestead is evaluating your available area and resources. A detailed assessment of your available land, water sources, climate, soil quality, existing structures, time, and personal talents are all part of this process. You may create a practical and successful plan to optimize the potential of your homestead by being aware of these variables. This section will go into great length about how to evaluate your space and resources, giving potential homesteaders a thorough how-to manual.

Making a thorough inspection of your property is the first step in evaluating your area. This entails calculating your land's size, determining its borders, and sketching out its features. You can better plan the layout of your homestead, including garden plots, animal enclosures, and utility areas if you are aware of the size and form of your property. Any existing buildings, like greenhouses, garages, or sheds, should be noted as well because they may be used for homesteading purposes.

Knowing the geography of your land is essential to space assessment. This entails determining elevations, slopes, and natural drainage patterns. Your property's topography has an impact on soil erosion, water flow, and microclimates. For example, gardens can be planted on areas with gradual slopes since they minimize soil erosion and offer appropriate drainage. On the other hand, terracing or other erosion management techniques can be necessary for steep slopes. Flat spaces are appropriate for greenhouses, livestock cages, and other constructions that need stable ground.

Every homestead needs water; thus, it's critical to evaluate your water supplies. This entails locating possible rainwater sources in addition to naturally occurring water sources such as ponds, streams, and groundwater. Planning irrigation systems, rainwater harvesting setups, and water storage solutions will be made easier if you are aware of the quantity and quality of water that is available on your land. Because periods of drought or intense rainfall might affect your homestead's water supply, it is crucial to take into account the seasonal unpredictability of water sources. It is also advised that you test your water for nutrients and impurities since this will help you make decisions about how to treat and use the water.

Another important consideration when evaluating your space and resources is the climate. The kinds of plants and animals that grow well on your property, as well as the duration and level of seasonal activity, are all determined by the climate in your area. Temperature, precipitation, humidity, and the dates of frost are important climate variables to take into account. Knowing the climate where you live will help you choose the right plants, arrange when to plant and harvest, and overcome weather-related obstacles. For instance, to protect your plants from frost in areas with harsh winters, you might need to make an investment in season extension strategies like cold frames, hoop buildings, or greenhouses.

Because soil quality has a direct impact on plant development and productivity, it is an essential component of successful homesteading. Finding out the fertility, pH, texture, and amount of organic matter in your soil requires doing a soil test. Independent labs or regional agricultural extension offices can conduct soil studies. Important information on nutrient imbalances and shortages will be revealed by the soil test findings. These can be remedied using soil amendments like compost, manure, lime, or mineral supplements. Adding organic matter, rotating crops, and reducing soil disturbance are all ongoing steps in enhancing soil health and fostering nutrient cycling and beneficial microbial activity.

The vegetation that already exists on your land can provide information not only about the quality of the soil but also about whether or not certain sections are suitable for gardening or other uses. Native plants and weeds can provide information on soil properties, moisture content, and sunshine availability based on their kinds and overall health. For instance, the existence of specific weeds may indicate a deficiency in nutrients or compacted soil. Finding places with a wide variety of healthy vegetation indicates ideal circumstances for crop growth.

The quantity of sunlight affects how productive your garden and other plant-based endeavors can be. Seeing how sunlight and shade are distributed over your property at different times of the day and in different seasons is part of doing a sunlight assessment. This will assist you in determining the ideal spots for plants that can withstand shadow, crops that thrive in the sun, and buildings that could produce more shade. This procedure can be aided by tools like shade analysis software or sun calculator. For most vegetable crops, it is best to provide your garden with six to eight hours of direct sunlight every day; however, fruit trees and perennials may require various amounts of light.

The buildings that are currently on your land may contain useful resources for your homestead. You can use sheds, garages, barns, and greenhouses for plant cultivation, animal housing, and storage. You can decide what repairs or changes are necessary to fit your homesteading activities by evaluating the functionality and state of these structures. In order to support your operations, you should also think about building new buildings like tool sheds, compost bins, or chicken coops.

When organizing your farm, personal abilities and time availability are critical resources to take into account. Evaluating your current proficiency in carpentry, gardening, animal husbandry, and other related fields will assist you in determining your areas of strength and growth. A wide range of skills are frequently needed for homesteading, so it's important to be honest about your limitations in order to set reasonable goals and keep yourself from being overwhelmed. It will boost your confidence and competency to devote time to acquiring new skills through books, online courses, workshops, and practical experience.

Another essential component of effective homesteading is time management. Think about your availability for homesteading duties on a daily, weekly, and seasonal basis. It takes careful planning and priority to manage homesteading activities with job, family, and other obligations. Making a plan or calendar for chores related to planting, harvesting, caring for animals, and maintenance will keep you organized and guarantee that important things don't go overlooked. Including time for relaxation and self-care is equally crucial if you want to avoid burnout and keep your general well-being.

When evaluating your space and resources, your financial resources should also be taken into account. Making a budget for your homesteading endeavors will assist you in setting aside money for necessary supplies, tools, and infrastructure upgrades. Begin by enumerating the early expenses for tools, building materials, soil amendments, seeds, and plants. Don't forget to account for recurring costs like maintenance, water and power, and animal feed. It's a good idea to budget for unforeseen expenses and contingencies as well. Keeping an eye on your spending plan will enable you to pick investments that will yield the most returns for your homestead and make well-informed decisions.

Creating a thorough homestead plan is the next step once you have evaluated your available space and resources. This plan should include your objectives, a schedule for completing the various initiatives, and a list of priorities. As you acquire knowledge and confidence, progressively increase the size of your efforts by starting with small, doable projects that fit your resources and skill set. Establishing both short- and long-term goals will give you a road map for realizing your vision and keep you motivated and focused.

For instance, your immediate objectives can be building a chicken coop, installing a composting system, and starting a small vegetable garden. Long-term objectives can be growing your garden, planting fruit trees, installing a rainwater collection system, and looking into alternative energy sources. Your homestead will continue to grow and change if you routinely examine and modify your strategy in light of your accomplishments and evolving situation.

Joining online or local communities and networking with other homesteaders can be a great way to get support and ideas. These networks provide chances to collaborate on initiatives and events and to exchange information, resources, and experiences. Joining a group of people who share your interests can improve your education, offer helpful advice, and foster a feeling of togetherness and purpose.

To sum up, the first step to a successful backyard homestead is evaluating your available space and resources. You can create a practical and efficient strategy to optimize the potential of your homestead by carefully assessing the size, topography, water sources, climate, soil quality, and existing structures on your property, as well as your personal abilities, time availability, and financial resources. You will be able to establish realistic goals, make well-informed decisions, and design a rewarding and sustainable homesteading lifestyle with the help of this thorough assessment. As you set out on your adventure, keep in mind that homesteading is an ongoing learning process that calls for tolerance, flexibility, and a dedication to coexist peacefully with the natural world.

Planning and goal setting

Setting goals and making a plan are essential components of any successful project, especially when it comes to backyard homesteading. This all-inclusive process entails planning your homestead's future, setting attainable goals, and developing a methodical strategy to get there. Through meticulous planning and goal-setting, homesteaders may guarantee that their endeavors are targeted, effective, and eventually profitable. This section will examine the complexities of goal-setting and planning for backyard homesteading, offering a thorough analysis of the actions, factors, and approaches required to establish a successful and long-lasting homestead.

The first step in the trip is to envision your homestead. Your values, hobbies, and preferred way of living should all be reflected in this vision, which should also represent your long-term goals. Whether your goal is to become totally self-sufficient, develop a close relationship with the natural world, or just savor the advantages of fresh, locally-grown food, having a clear vision gives you focus and inspiration. This vision will act as a beacon of hope for you while you plan and establish your goals, assisting you in navigating the possibilities and obstacles that present themselves along the road.

After you have a clear vision, you need to take a close look at your current circumstances. This entails assessing your assets, capabilities, time constraints, and abilities. Knowing the advantages and disadvantages of your current configuration will help you plan and create goals that are both reasonable and doable. Examining your property's dimensions, composition, water availability, soil condition, and existing structures are all part of the assessment process. Analyzing your resources also entails taking into account your financial plan, your tools and equipment, and possible partners or sources of support.

Once you have a clear idea of where you are beginning, you can start to create SMART goals—specific, measurable, achievable, relevant, and time-bound. These objectives ought to be customized to your particular situation and in line with your vision. If your objective is to create a profitable and diverse garden, for instance, a SMART goal could be to plant ten different crops in a vegetable garden before the next growing season. Establishing a vegetable garden with ten crops is a specific goal that is time-bound (within the next growing season), relevant (aligned with your vision of a productive garden), measurable (based on the number of crops planted and harvested), achievable (based on your assessment of resources and skills), and measurable.

It is imperative that you divide your goals into smaller, more achievable activities in order to aid in their accomplishment. Progress can be made steadily and overwhelm can be avoided with this methodical approach. For example, the objective of starting a vegetable garden can be broken down into smaller tasks like planning the garden's layout, getting the soil ready, choosing and buying seeds, planting, watering, and caring for the garden. Keeping yourself organized and motivated can be achieved by making a thorough schedule for these tasks that includes deadlines and milestones. Additionally, it gives you a sense of success as you finish each assignment and get closer to your main objective.

In backyard homesteading, resource management is an essential component of planning and goal-setting. This entails planning your spending, managing your time well, and utilizing the supplies and equipment that are at your disposal. For your household tasks, creating a budget will assist you in setting spending priorities and preventing needless debt. It's critical to account for both one-time expenditures (such as buying supplies, seeds, and equipment) and continuing costs (like maintenance, water, and power). Maintaining a close eye on your expenditures and periodically assessing your spending plan can guarantee that your money is spent sensibly and sustainably.

Effective time management is crucial, particularly for individuals who have to juggle work, family, and other obligations in addition to homesteading. To stay organized and avoid duties building up, make weekly and monthly schedules outlining your homesteading activities and tasks. It's critical to estimate each task's time realistically and to leave room for unforeseen circumstances or difficulties. Recognizing the seasonality of various homesteading tasks, like planting, harvesting, and canning, will also help you schedule your time more efficiently and maximize each season.

Assessing and improving your talents is essential for effective homesteading, in addition to having the necessary time and money. This entails determining the skills you currently have and which ones you must acquire in order to reach your objectives. For example, you may need to learn about coop construction, egg gathering, and chicken care if your objective is to keep hens for eggs. Fortunately, there are lots of tools out there to help you learn new skills, like neighborhood groups, online courses, books, and workshops. Devoting time towards skill enhancement will boost your self-assurance and proficiency, allowing you to take on a greater variety of homesteading tasks.

Engaging in the community and networking are important tactics for reaching your homesteading objectives. Developing ties with nearby farmers, gardeners, and other homesteaders can be a source of inspiration, support, and useful help. Participating in online forums, homesteading groups, or gardening clubs in your community can help to foster collaboration and knowledge exchange. Engaging in community events like farmers' markets, seed exchanges, and workshops can also help you meet new people and broaden your network by exposing you to fresh concepts and methods. Strong ties to the community can also foster a spirit of unity and support for one another, which is especially helpful in trying times.

Planning and goal-setting need you to keep track of and assess your progress. You may keep yourself on track and make any adjustments by often assessing your objectives, assignments, and deadlines. This process entails evaluating what is going well, pinpointing areas that require development, and acknowledging your accomplishments. One useful tool for measuring your progress and recording your experiences with homesteading is to keep a journal or log of your activities. This document can be a source of inspiration and encouragement as well as insightful information for future planning.

Adaptability and flexibility are necessary traits for prosperous homesteaders. Unexpected possibilities and problems will inevitably present themselves even with meticulous planning. Your ability to adapt and be flexible when faced with unforeseen events will enable you to deal with uncertainty with fortitude and inventiveness. For example, you might need to replant with a different variety or use new pest control techniques if a specific crop fails because of pests or weather. In a similar vein, you could need to reevaluate and modify your objectives to account for any new assets if the chance to obtain more land or resources presents itself.

Establishing goals and planning with sustainability in mind is essential to building a resilient and eco-friendly household. This entails implementing methods that support ecological health, minimize waste, and conserve resources. Composting and mulching, for instance, can increase soil fertility and lessen the demand for chemical fertilizers. You may save water and lessen your need for municipal supplies by implementing rainwater collection systems and effective irrigation techniques. By including renewable energy sources like wind turbines or solar panels, you can improve your energy independence and lessen your carbon footprint. Making sustainability a top priority helps the environment and builds a more self-sufficient farm.

For those who want to establish a long-lasting and sustainable way of life, homesteaders must take long-term planning into account. This entails planning for your homestead's long-term expansion and development in addition to your current objectives. For instance, planting perennial crops and fruit trees can improve soil health and offer long-term food sources. Building structures like root cellars, barns, and greenhouses will increase your ability to produce and store food. A more resilient and sustainable homestead can also be built over time by making investments in soil improvement techniques, water conservation strategies, and renewable energy systems.

Fostering a cooperative and encouraging homesteading environment requires involving family and household members in the planning and goal-setting process. Talking with your family about your vision and objectives will help focus everyone's efforts and establish a feeling of purpose. Engaging kids and other family members in homesteading pursuits can enhance educational opportunities and build familial relationships. It is also possible to more fairly divide the burden and guarantee that each person contributes to the homestead's success by giving age-appropriate chores and responsibilities.

Planning and goal-setting for homesteading should take health and well-being into account. Setting aside time for activities that advance your mental and physical well-being can improve your life in general and help you reach your objectives. This entails establishing areas for rest and recreation on your homestead, keeping a healthy, balanced diet with vegetables grown locally, and getting frequent exercise from gardening and outdoor activities. Moreover, practicing stress-reduction methods like yoga, meditation, or mindfulness might support you in keeping an optimistic and resilient outlook.

Planning and goal-setting also require careful consideration of where to get and how to manage the inputs for your homestead. Purchasing the seeds, plants, animals, feed, soil additives, equipment, and supplies you need for your projects are all included in this. The quality and sustainability of your inputs can be improved by creating a network of trustworthy suppliers and taking into account regional and sustainable solutions. For example, you can guarantee genetic diversity and adaptation by buying heirloom seeds from reliable vendors or from nearby seed exchanges. In a similar vein, purchasing livestock from sustainable and ethical breeders will enhance your animals' well-being and output.

An essential part of planning for homesteading is risk management. You can preserve your homestead and overcome obstacles by recognizing possible dangers and creating plans to reduce them. Natural disasters, illnesses and pests, equipment malfunctions, and financial setbacks are a few examples of risks. You may be more resilient and prepared by creating backup plans, such as keeping an emergency fund, having backup water supplies, and putting pest control techniques into practice. Purchasing insurance for your equipment, animals, and property can also offer financial security in the event of unanticipated circumstances.

The process of planning and creating goals is inextricably linked to education and ongoing learning. You may improve your knowledge and abilities by keeping up with the latest advancements in sustainable technologies, agricultural research, and homesteading techniques. Reading books and articles, taking online courses, and attending webinars, conferences, and workshops can all offer insightful and inspiring experiences. You can also shorten your learning curve and steer clear of typical traps by reading about other homesteaders' experiences and asking more seasoned people to advise you.

One crucial factor to take into account in contemporary homesteading is the use of technology. Although conventional methods and abilities are essential, using the right technologies can increase effectiveness and output. For instance, you can design and maximize the layout of your garden by using garden planning software. Automated irrigation system installation can guarantee regular watering and save time. You can cut your energy expenses and dependency on fossil fuels by using solar-powered tools and equipment. You can build a more productive and long-lasting homestead by striking a balance between traditional and contemporary methods.

Planning and goal-setting achievement requires meticulous record-keeping and documentation. Planning ahead can benefit greatly from having thorough records of all your homesteading operations, including planting schedules, crop yields, animal health, costs, and maintenance duties. You can find trends in this documentation, monitor developments, and make wise choices. Keeping up a blog or journal on your homestead can also act as a personal chronicle of your journey, documenting your accomplishments, setbacks, and experiences.

Planning your homestead includes being aware of zoning laws and municipal regulations. You may avoid potential problems and guarantee that your actions are compliant by being aware of the legal requirements and constraints in your area. This could entail getting licenses for specific buildings, abiding by livestock zoning laws, and using water and waste management best practices. You can manage these rules and steer clear of legal concerns by speaking with local officials and getting guidance from other homesteaders in your area.

To sum up, goal-setting and planning are essential steps toward building a prosperous and long-lasting backyard homestead. You may create a systematic and successful strategy by establishing SMART goals, breaking them down into achievable tasks, making a thorough assessment of your existing condition, and developing a clear vision. Organizing your resources, developing your abilities, becoming involved in your community, and using sustainability concepts will all help you in your endeavors. You may overcome obstacles and realize your long-term goals by routinely reviewing and modifying your strategy, being adaptive and flexible, and placing a high priority on your health and well-being. You may build a robust and successful homestead that increases your quality of life and reflects your values with careful planning and persistent work.

CHAPTER II

Soil Preparation and Gardening Basics

Techniques for improving soil fertility

A key element of productive farming, gardening, and land management is healthy soil. It speaks to the capacity of the soil to give plants the necessary nutrients at sufficient levels for growth and development. Many methods are used to improve soil structure, nutrient content, and biological activity in order to increase soil fertility. The application of organic amendments, crop rotation, cover crops, green manures, composting, mulching, biochar application, soil testing and amendments, microbial inoculants, and sustainable land management strategies are just a few of the methods for increasing soil fertility that are thoroughly examined in this section.

One of the best ways to increase soil fertility is with organic additions. Compost, manure, and green waste are examples of amendments that contribute organic matter to the soil, improving its structure, ability to hold water, and nutrient content. Plant growth requires a steady supply of nutrients, which organic matter offers. For instance, compost is a great source of micronutrients like calcium, magnesium, and sulfur, in addition to macronutrients like potassium, phosphorus, and nitrogen. Additionally, it has helpful microbes that aid in the breakdown of organic matter and release nutrients in a form that is absorbed by plants. Manure gradually increases the fertility of the soil by adding a large amount of organic matter and nutrients when it is properly composted. Green waste, such as grass clippings and leaves, can either be added to the soil directly or composted to increase the amount of organic matter and fertility.

A classic agricultural technique called crop rotation is planting several crop varieties in the same spot over the course of several growing seasons. By stopping the loss of particular nutrients and lowering the accumulation of pests and illnesses linked to ongoing monoculture, this method aids in the management of soil fertility. The root systems and nutrient requirements of different crops vary, which has an impact on soil health in diverse ways. Legumes, like beans and peas, for instance, have the capacity to fix atmospheric nitrogen thanks to symbiotic partnerships with bacteria that fix nitrogen in their root nodules. Increasing soil nitrogen levels and lowering the need for synthetic fertilizers can be achieved by planting legumes in rotation with other crops, such as grains or vegetables. While shallow-rooted crops can make use of nutrients close to the soil surface, deep-rooted crops, such as alfalfa or sunflowers, can aid in fracturing compacted soil layers and enhancing soil structure.

A sustainable agricultural method called cover cropping is planting certain crops or cover crops when the surrounding soil would otherwise be bare. Clover, vetch, rye, and buckwheat are examples of cover crops that offer a variety of advantages, such as reducing erosion, cycling nutrients, suppressing weeds, and supplying organic matter. After being cut down and replanted, cover crops break down and release nutrients into the soil, improving soil fertility. Because legumes can fix nitrogen, they are particularly good at increasing the amount of nitrogen in the soil. Through the creation of channels that facilitate water infiltration and root penetration for succeeding crops, cover crop roots also help to improve the structure of the soil. Furthermore, cover crops support biodiversity and soil microbial activity, both of which are critical to soil health and nutrient cycling.

A particular kind of cover crop known as "green manures" is cultivated mainly to be tilled under and mixed into the soil while still green and developing. This procedure improves the fertility of the soil by quickly adding organic matter and nutrients. Alfalfa, mustard, and clover are examples of green manures that are chosen for their rapid growth and high biomass production. Green manures break down fast in the soil, releasing nutrients and strengthening the soil's structure. When significant nutrient demand is needed, such as in gardens or intensive agricultural systems, this strategy works very well to improve soil fertility. By keeping the ground covered during fallow periods, green manures also aid in the suppression of weeds and the reduction of soil erosion.

By converting organic waste into useful soil nutrients, composting is a crucial method for increasing soil fertility. Through microbial activity, composting involves the controlled breakdown of organic materials including kitchen scraps, yard waste, and manure. Rich in nutrients and advantageous microbes, the compost that results improves the fertility and structure of the soil. A balanced mixture of carbon-rich (browns) and nitrogen-rich (greens) materials, as well as sufficient moisture and aeration, are necessary for successful composting. To enhance the quality of the soil, finished compost can be added to planting beds, used in potting mixes, or spread over the soil as a mulch. In addition to offering vital nutrients, compost enhances soil microbial activity, water retention, texture, and texture, all of which are conducive to plant growth.

Another useful method for enhancing soil fertility and preserving moisture is mulching. A layer of organic or inorganic material called mulch is spread over the soil's surface to control temperature, prevent erosion, stifle weeds, and hold onto moisture. Straw, wood chips, leaves, and grass clippings are examples of organic mulches that progressively break down and enrich the soil with organic matter, improving its fertility. Mulching reduces the need for regular watering and stops nutrient leaching by helping to maintain constant soil moisture levels. Additionally, it balances the temperature of the soil, shielding plant roots from intense heat or cold. Over time, organic mulches improve soil fertility by releasing nutrients into the soil that plants may use.

The process of applying biochar to soil is known as biochar application. Biochar is a type of charcoal that is made by pyrolyzing organic materials. Because of its porous nature, biochar improves microbial activity, water retention, and nutrient availability in soil, hence increasing soil fertility. It improves soil health by acting as a long-term nutrient reservoir, hence decreasing nutrient leaching. Additionally, by enhancing porosity and decreasing bulk density, which encourages root development and water infiltration, biochar enhances soil structure. Additionally, by storing carbon in the soil for extended periods of time, biochar can aid in reducing the effects of climate change. The sustainability and fertility of soil can be greatly increased by using biochar in soil management techniques.

To better understand and enhance soil fertility, soil tests and amendments are essential. Tests on the soil can reveal important details about the pH, organic matter content, nutrient levels, and other characteristics that affect plant growth. These tests can be carried out at home with testing kits, in labs, or through agricultural extension agencies. Appropriate amendments can be used to rectify pH imbalances and nutrient deficits based on the findings of soil tests. Lime (to raise soil pH), sulfur (to drop pH), and some fertilizers (to replace insufficient nutrients) are common soil additives. Frequent monitoring and testing of the soil enable focused soil management techniques that improve fertility and well-informed decision-making.

Microbial inoculants are soil additives that are enriched with advantageous microorganisms, like fungi and bacteria, that support plant health and soil fertility. These microbes are involved in the decomposition of organic materials, the cycling of nutrients, and the suppression of disease. As an illustration, mycorrhizal fungi develop symbiotic partnerships with plant roots to improve soil structure and nutrient uptake, especially phosphorus. The amount of nitrogen in the soil is increased by nitrogen-fixing bacteria, such as Rhizobium species, which transform atmospheric nitrogen into a form that plants can use. By increasing microbial activity and nutrient availability, microbial inoculants applied to soil or seeds can improve soil fertility. These inoculants are especially helpful in places with low microbial diversity or damaged soils.

Using sustainable land management techniques is crucial to preserving and enhancing soil fertility over time. Reducing soil disturbance, cutting back on chemical inputs, and fostering biodiversity are some of these techniques. Soil structure and organic matter content can be preserved by using no-till or reduced-till farming techniques, which minimize soil disturbance. Reduced use of synthetic pesticides and fertilizers lowers the chance of soil contamination and nutritional imbalances. Soil health and resilience are improved by promoting biodiversity through agroforestry, cover crops, and varied crop rotations. In addition to preserving natural resources like soil and water, sustainable land management techniques include ecological concepts in gardening and farming activities.

Agroforestry is a sustainable method of managing land that creates a mutually beneficial system by combining crops, livestock, and trees and shrubs. Agroforestry systems increase the amount of organic matter in the soil, decrease erosion, and improve nutrient cycling. In agroforestry systems, trees and shrubs supply root biomass and leaf litter, which break down and add nutrients to the soil. Additionally, their deep roots aid in recycling nutrients from deeper soil layers so that crops with shallow roots may get them. Agroforestry methods also produce microclimates that shield soil from harsh weather and improve the general health of ecosystems. Agroforestry techniques have the potential to greatly increase soil sustainability and fertility.

Soil fertility can also be increased by including livestock in farming operations. Because their dung provides nutrients and organic matter to the soil, livestock—including cattle, sheep, and poultry—helps maintain the fertility of the soil. In managed grazing systems, waste is distributed uniformly, and overgrazing is avoided by rotating cattle among several pasture areas. Livestock tramples the soil, incorporating organic matter into the soil and enhancing its fertility and structure, including animals in crop production, resulting in a closed-loop system that recycles nutrients, lowering the demand for outside inputs and improving soil health.

Soil fertility can be greatly increased by combining no-till or reduced-till farming techniques with the use of green manures and cover crops. Farming practices such as reduced-till and no-till preserve soil structure and organic content by minimizing soil disturbance. The synergistic effect of cover crops, green manures, and other measures enhances the fertility and health of the soil. Reduced-till techniques preserve soil structure and microbial activity, while cover crops and green manures enrich the soil with organic matter and nutrients. By enhancing soil fertility, water retention, and resilience, this combination develops an agricultural system that is both productive and sustainable.

In order to prevent erosion and preserve soil fertility, conservation tillage techniques must be used. Conservation tillage, which includes mulch-till, strip-till, and no-till methods, disturbs the soil as little as possible while maintaining the organic matter and soil structure. By preserving ground cover and shielding the soil's surface from wind and water erosion, these activities lessen erosion. In addition to increasing microbial activity and soil moisture retention, conservation tillage raises soil fertility. Farmers can sustain soil health and productivity over time by lowering the frequency and intensity of tillage.

Water management is essential to plant health and productivity and is strongly correlated with soil fertility. Adequate watering of plants is ensured by good irrigation techniques, which also prevent waterlogging and nutrient leaching.

Adequate watering of plants is ensured by good irrigation techniques, which also prevent waterlogging and nutrient leaching. Water loss and runoff are reduced when using efficient irrigation techniques like drip irrigation and soaker hoses, which supply water straight to the root zone. By increasing the soil's ability to store water, mulching, and adding organic matter, you can cut down on watering frequency and preserve water resources. Using water-saving strategies, such as collecting rainfall and utilizing graywater, increases the amount of water available to plants and lessens their dependency on outside water sources.

Optimizing nutrient availability and encouraging strong plant growth depends on controlling the pH of the soil. The availability and solubility of vital nutrients are influenced by the pH of the soil. The pH range of 6.0 to 7.5 is slightly acidic to neutral, as preferred by most plants. By adding soil amendments, such as sulfur, to drop pH or lime to raise pH, one can alter the pH of the soil, which can enhance nutrient uptake and soil fertility overall. To keep an eye on pH levels and make well-informed decisions on pH adjustments based on crop requirements and soil conditions, regular soil testing is required.

Maintaining optimum soil fertility requires incorporating nutrient management techniques. Applying fertilizers and soil amendments in amounts and timings that correspond with plant nutrition demand is known as nutrient management. Although synthetic fertilizers give plants easily accessible nutrients, if applied incorrectly, they can cause nutritional imbalances and soil damage. Compost and manure are examples of organic fertilizers that gradually release nutrients while enhancing soil microbial activity and structure. Using integrated nutrient management techniques, such as crop rotation, cover crops, and soil testing, helps minimize environmental effects and maximize fertilizer usage efficiency.

In order to maintain soil fertility and productivity, soil erosion control is crucial. Erosion is the process by which dirt is carried away from the surface of the ground by wind, water, or other factors, resulting in the loss of nutrients and important topsoil. Soil stabilization and erosion prevention can be achieved by putting erosion control techniques, including terracing, contour plowing, and planting vegetative buffers, into place. By decreasing surface runoff and improving soil structure, cover crops and permanent vegetation cover prevent soil erosion. Farmers and gardeners may maintain soil fertility and sustainable land use by limiting soil disturbance and maintaining ground cover.

Conservation of biodiversity and soil fertility are enhanced when agroecological concepts are included in gardening and agricultural practices. Agroecology focuses on ecological processes to improve soil health, biodiversity, and nutrient cycling in order to increase agricultural resilience and productivity. Agroforestry, polyculture, and mixed cropping systems are examples of practices that increase plant species diversity and offer a variety of ecological advantages. Enhanced organic matter management, decreased chemical inputs, and biodiversity promotion are the main goals of agroecological farming practices, which place a high priority on soil health and fertility. Agroecological techniques promote long-term soil fertility and ecosystem resilience by imitating natural ecosystems.

Maintaining soil fertility in the face of shifting climatic conditions requires building soil resilience to climate change impacts. The health and productivity of soil are threatened by climate change, which includes changes in temperature, precipitation patterns, and extreme weather events. Mitigating the effects of climate change on soil fertility can be achieved by implementing adaptive soil management techniques, such as strengthening soil structure, optimizing water management, and encouraging carbon sequestration. By enhancing water retention, lowering erosion, and promoting microbial activity, techniques including cover crops, conservation tillage, and agroforestry improve soil resilience. Farmers and gardeners can maintain agricultural productivity and adjust to climatic variability by enhancing soil resilience.

Monitoring soil health is crucial for determining the potential for improvement and evaluating soil fertility. In order to determine the soil's ability to sustain plant development and ecosystem activities, soil health evaluations examine the physical, chemical, and biological characteristics of the soil. Decisions about soil management are aided by the quantitative information that soil tests provide on nutrient levels, pH, organic matter content, and microbial activity. It is possible to follow changes in soil fertility over time by keeping an eye on indicators of soil health such as earthworm populations, water infiltration rate, and soil structure. Sustainable soil fertility and productivity optimization can be achieved by farmers and gardeners via the implementation of proactive monitoring and management strategies.

In summary, enhancing soil structure, nutrient availability, and biological activity all work together to improve soil fertility. Compost and manure are examples of organic amendments that enrich the soil with nutrients and organic matter, enhancing its fertility and overall health. Crop rotation, cover crops, and green manures are examples of practices that improve soil structure, restore nutrients, and inhibit pests and diseases. Mulching and composting increase soil fertility by preserving moisture, regulating temperature, and recycling nutrients. Applying biochar and microbial inoculants helps to increase soil microbial activity and retain nutrients. Long-term soil fertility and resilience are supported by sustainable land management techniques, such as agroecological methods, water management, and erosion control. Farmers and gardeners can sustain healthy, productive soils that promote sustainable agriculture and ecosystem health by putting these strategies and practices into practice.

Composting and vermiculture

Vermiculture and composting are two related, environmentally friendly waste management techniques that have drawn a lot of attention for their capacity to recycle organic waste into beneficial soil additions. These techniques improve soil health, encourage plant development, and lessen environmental contamination, in addition to assisting in the reduction of the amount of garbage dumped in landfills. This section explores the nuances of vermiculture and composting, highlighting its advantages, disadvantages, and critical role in promoting a sustainable future.

The aerobic, regulated breakdown of organic waste by microorganisms is known as composting. Organic waste, including leftover food, yard debris, and agricultural residues, is converted into nutrient-rich compost via this organic waste conversion process. The mesophilic, thermophilic, and maturation phases are the three basic stages of the composting process. Mesophilic bacteria break and degrade simple organic molecules at moderate temperatures during the few-day mesophilic stage. The next stage is thermophilic, which is distinguished by high temperatures brought on by the activities of fungi and bacteria that are thermophilic. Several weeks to several months can pass during this phase, depending on the composting technique and materials utilized. Pathogens are eliminated at this step, and more intricate organic materials like cellulose and lignin break down. At last, the compost cools down, and mesophilic bacteria take over once more, further stabilizing the compost and preparing it for use. This is known as the maturation phase.

Temperature, moisture content, aeration, and the feedstock's carbon-to-nitrogen (C: N) ratio are the variables that affect the composting process. Retaining an ideal temperature range of 40–60°C is essential for effective pathogen elimination and breakdown. The ideal moisture content range is 40–60% since too much moisture can cause anaerobic conditions, and too little moisture can inhibit microbial activity. In order to supply oxygen to aerobic bacteria and stop the growth of bad smells, aeration is essential. Proper aeration can be maintained by stirring the compost pile on a regular basis. The optimal C: N ratio for microorganisms is between 25:1 and 30:1, which guarantees a well-balanced diet. Straw and wood chips are high in carbon and offer energy; food scraps and grass clippings, on the other hand, are rich in nitrogen and provide proteins needed for microbial growth.

There are several social, economic, and environmental advantages to composting. In terms of the environment, it lessens the quantity of organic waste that ends up in landfills, which lowers methane emissions—a strong greenhouse gas. Additionally, composting recycles nutrients, lessening nutrient runoff that can contaminate water and decreasing the demand for synthetic fertilizers. Compost also enhances soil fertility, water retention, and structure, all of which support strong plant growth. Composting is a cost-effective way for communities to reduce their waste disposal expenses and make money from compost sales. By generating jobs in composting plants and associated businesses, it also boosts local economies. Composting promotes environmental stewardship and sustainable waste management techniques by educating and engaging the community.

Composting presents a number of difficulties despite its advantages. One of the biggest problems is when non-compostable items, like metals and plastics, contaminate the feedstock. This can slow down the composting process and lower the quality of the finished product. To guarantee appropriate waste segregation at the source, public awareness and education are crucial. Managing pests and smells presents another difficulty, although these can be avoided with the use of biofilters or other odor-management technology, appropriate aeration, and moisture control. Furthermore, composting facilities need a lot of equipment and space, which can be limited in urban settings. Decentralized composting solutions, on the other hand, such as home and community composting, can solve this problem.

Vermiculture, also known as vermicomposting, is a particular kind of composting in which organic waste is broken down by earthworms. The most often utilized earthworm species in vermiculture are red wigglers or Eisenia fetida. Comparing vermiculture to conventional composting techniques reveals a number of benefits. Because they break down organic materials and increase microbial activity in their intestines, earthworms expedite the decomposition process. The resultant vermicompost, also known as worm castings, is a great soil supplement because it is full of humic materials, microbes, and nutrients.

The method of vermiculture entails creating an atmosphere that is conducive to earthworm growth. A vermiculture system usually consists of a bin or container that is filled with bedding material, which gives the earthworms a place to live. Examples of this bedding material include cardboard, coconut coir, and shredded paper. Add organic garbage to the bin to feed the earthworms, such as leftover fruit and vegetable pieces, coffee grounds, and plant trimmings. Maintaining the ideal environment for the earthworms, which includes a temperature range of 15–25°C, a moisture content of 70–80%, and adequate aeration, is crucial. The productivity and well-being of the earthworm population depend on routine feeding and bin monitoring.

Vermiculture has a number of special advantages. In terms of microbial variety and nutrient content, vermicompost is better than regular compost. Higher concentrations of micronutrients like calcium, magnesium, and iron are present, along with important elements like potassium, phosphorus, and nitrogen. Additionally, vermicompost improves soil aeration, water retention, and structure, which encourages strong root development and raises crop yields. Furthermore, vermiculture is an inexpensive, low-tech technique that is simple to apply on a small scale, making it available to homes, community gardens, and schools. Additionally, it offers a teaching tool for biology, ecology, and sustainable behaviors to be used with both adults and children.

Vermiculture does, however, also encounter certain difficulties. Since earthworms are sensitive to changes in pH, moisture content, and temperature, one of the biggest obstacles is keeping their environment at ideal levels. The vermiculture process can be hampered, and earthworms can be damaged by extreme temperatures, high levels of wetness, or acidic environments. To avoid these problems, proper administration and oversight are necessary. Potential pests like fruit flies and mites provide another difficulty, but these may be managed with proper hygiene and routine bin maintenance. Furthermore, vermiculture systems may not be appropriate for managing substantial volumes of organic waste because of their restricted capacity. This restriction can be somewhat overcome by combining vermiculture with other composting techniques or waste management strategies.

Composting and vermiculture combined provide a comprehensive method for managing organic waste. Combining these techniques will allow you to take full advantage of their complimentary advantages. Vermiculture can be utilized for high-value applications, such as creating nutrient-rich vermicompost for gardening and horticulture, whereas composting can be used to treat bigger amounts of organic waste and produce bulk compost. Furthermore, to improve soil health, biodiversity, and ecosystem resilience, composting and vermiculture can be used with other sustainable techniques, including crop rotation, cover crops, and agroforestry.

The promotion of composting and vermiculture is greatly aided by government regulations and neighborhood projects. Policies, grants, and instructional initiatives are some of the ways that governments can encourage these behaviors. Vermiculture and composting can become more popular through laws that, for instance, require the separation of organic waste at the source, finance composting facilities, and encourage the use of compost in landscaping and agriculture. Community-based programs that educate and encourage a sustainable culture include demonstration gardens, compost cooperatives, and composting workshops. To create an inclusive and strong organic waste management system, cooperation between enterprises, non-profits, municipalities, and educational institutions is crucial.

Vermiculture and composting are evolving as a result of technological improvements. Composting operations are becoming more efficient and scalable thanks to technological advancements in composting, such as temperature control, automated aeration, and in-vessel composting systems. Improved vermiculture methods, such as selective breeding for more robust and fruitful earthworm strains, are being made possible by research on the biology and behavior of earthworms. Furthermore, digital platforms and tools like online markets and smartphone apps are making it easier for consumers and practitioners to share knowledge and collect and distribute compost and garbage.

Composting and vermiculture have advantages for the environment, economy, and society that go beyond waste management. Through improving soil carbon sequestration and lowering greenhouse gas emissions from landfills, these methods help mitigate the effects of climate change. By boosting organic agricultural methods, lowering dependency on chemical fertilizers, and increasing soil fertility, they also assist sustainable agriculture. Furthermore, by offering premium soil additives for community gardens, small-scale farms, and urban and peri-urban agriculture, composting and vermiculture support local food systems. As people work together to manage organic waste and nourish the soil, they also promote a sense of community and environmental care.

The advancement of vermiculture and composting depends on outreach and education. Educating people about the advantages and methods of these activities can inspire people to take action on their own and in their communities. When it comes to imparting knowledge and expertise on composting and vermiculture, educational institutions, universities, and extension services might be essential. The adoption of sustainable waste management strategies can be accelerated by incorporating these issues into curricula, hosting interactive seminars, and encouraging research and innovation. Moreover, social media, community activities, and public awareness campaigns can captivate a wider audience and motivate group efforts toward sustainability.

Vermiculture and composting are essential elements of a sustainable waste management system, to sum up. Waste reduction, nutrient recycling, soil enhancement, and climate change mitigation are just a few advantages of these activities. Composting and vermiculture can be successfully adopted despite their obstacles if they are managed properly, educated, and given assistance by government regulations and neighborhood projects. Vermiculture and composting can have a greater impact and support resilient, regenerative ecosystems when combined with other sustainable activities. Adopting composting and vermiculture will be crucial as we progress toward a more sustainable future because they will support the development of a circular economy, healthy soils, and an environmentally conscious culture.

Planning your garden layout

Creating a useful, beautiful, and sustainable garden area involves careful consideration of many different aspects, which is a complicated process in garden layout planning. Whether you are a newbie or an expert gardener, making the most of your space, maintaining the health and yield of your plants, and fostering a peaceful atmosphere all depend on careful planning. This section provides thorough assistance to help you establish a successful garden layout by examining the essential elements of garden layout planning, such as site evaluation, design principles, plant selection, soil preparation, water management, and maintenance strategies.

To begin designing a garden layout, a thorough site survey must be carried out. It is essential to comprehend the distinct features of your garden location in order to make well-informed selections regarding plant selection and design. First, note the terrain, kind of soil, amount of sunlight, and flora that is currently present on the site. Make a note of the garden's microclimates, drainage patterns, and slopes. Topography affects soil erosion and water runoff, which can have an impact on plant development and garden stability. The structure, nutrient availability, and drainage capacity of the soil are all influenced by the kind of soil—clayey, loamy, or sandy. You can make the necessary soil amendments by testing the soil's pH and nutrient levels. Given that different plants have distinct light requirements, sunlight exposure is essential to the health of plants. Determine whether spots get full light, half shade, or full shade all day long. Suggestions on soil properties and microclimates can be found in the existing vegetation. For example, large trees can affect the amount of moisture in the soil and produce shaded areas. It is possible to create a garden that blends in with the natural surroundings by having an understanding of these site qualities.

The next stage is to create a garden design that fulfills your demands and expresses your vision once you have evaluated the land. Garden settings that are both aesthetically beautiful and utilitarian are created by following garden design principles: unity, balance, contrast, rhythm, and scale. To achieve unity, one must design a harmonious garden in which every component enhances the others. Consistent plant selections, color palettes, and design motifs can help achieve this. The distribution of visual weight in a garden is referred to as balance. To achieve symmetrical balance and a formal, structured appearance, reflect components on either side of a central axis. In contrast, asymmetrical balance creates a sense of stability by utilizing many parts, which feels more organic and carefree. By contrasting various textures, colors, and shapes, contrast enhances visual interest. Striking contrasts can be produced, for instance, by mixing coarse-textured and fine-textured plants or by contrasting bright flowers with dark foliage. Repetition of elements is used in rhythm to imply movement and continuity. Repetition of plant groups, colors, or garden elements can accomplish this. The proportion of garden items to one another and the total garden area is referred to as scale. It's important to make sure that structures, paths, and plants are all proportioned adequately to avoid the garden appearing small or empty.

Think about things like climate, soil type, availability of water, and growth patterns while choosing plants for your garden. To maintain the health and vigor of your plants, choose ones that are appropriate for the soil type and local climate. Since native plants are more suited to the area and require less care, they are frequently a suitable choice. To construct water-efficient zones, put plants with comparable requirements together and take into account the water requirements of the individual plants. A healthy and sustainable garden requires an understanding of plant growth characteristics, including mature size, growth rate, and seasonal variations. Give each plant adequate room to grow to its maximum size without crowding out other plants for resources. This will help you avoid overcrowding. To provide year-round visual appeal, take into account the seasonal interest of the plants, including their fruit production, color of the foliage, and blooming seasons. A combination of shrubs, trees, perennials, and annuals may add structure, variation, and year-round interest to your landscape.

A vital component of designing a garden layout that directly affects plant productivity and health is soil preparation. Water, nutrients, and support are all provided by healthy soil for plant growth. Start by adding organic amendments, like as compost, aged manure, and leaf mold, to the soil to improve its fertility and structure. These substances improve the texture of the soil, make more nutrients available, and encourage healthy microbial activity. Adding organic matter also helps sandy soils retain more water and enhances drainage in heavy clay soils. Mulching is an additional useful technique for managing soil that helps control temperature, discourage weed growth, and preserve moisture. For the purpose of protecting the soil and lowering maintenance requirements, spread an organic mulch layer around plants, such as wood chips, straw, or shreds of leaves. Test the soil frequently to keep an eye on its pH and nutrient content, and make necessary adjustments to keep the ideal circumstances for plant growth.

When designing a garden layout, water management is quite important, especially in areas with scarce water supplies or erratic rainfall patterns. It is possible to guarantee that plants receive enough water without wasting resources by designing an effective irrigation system. Water can be delivered directly to plant roots via drip irrigation and soaker hoses, which reduces evaporation and runoff. These systems can be linked to moisture sensors or timers to automate watering schedules and cut down on labor-intensive tasks. Rainwater collection and utilization is an additional sustainable approach that can be employed to meet irrigation requirements. To collect runoff from roofs and other surfaces and store it for use during dry spells, install rain barrels or a rainwater harvesting system. Water usage can be further decreased by implementing xeriscaping techniques and drought-tolerant plant selections. Choose plants that don't require much water and arrange them in groups according to how much moisture they demand. Use swales and curves in the garden's design to direct and hold onto rainwater so that it can seep into the ground and replenish groundwater.

The arrangement of garden elements and buildings is crucial to the overall look and usefulness of the garden, in addition to plant selection and soil preparation. To create a harmonious and easily accessible garden environment, the placement of pathways, seating places, plant beds, and focal points should be carefully considered. Pathways direct traffic and give the landscape structure. To make sure routes are both functional and visually beautiful, take into account their breadth, material, and alignment. Popular materials like wood, stone, brick, and gravel can go well with a variety of garden designs. Patios, pergolas, and benches are examples of seating places that provide locations for leisure and relaxation. Arrange sitting sections to benefit from views, shade, and seclusion, resulting in cozy places to relax and commune with the natural world. Different plantings can be accommodated and visual interest can be created by designing garden beds in a variety of forms and sizes. Raised beds are very helpful for increasing accessibility and soil drainage. Garden focal points, such as sculptures, water features, and specimen plants, grab the eye and give the space personality. Strategically position focus points to improve vistas and foster a sense of exploration.

Planning a garden's layout must include maintenance techniques in order to guarantee the garden's long-term health and aesthetic appeal. Weeding, trimming, fertilizing, watering, and insect control are examples of routine maintenance chores. Making a routine for upkeep keeps the garden in top shape and keeps minor problems from turning into bigger ones. Watering should be done effectively and consistently, considering the needs of various plants as well as the climate. Watering deeply and sparingly promotes drought resistance and deep-root development. In order to keep things neat and avoid competition for resources, weeding is necessary. Hand weeding and mulching are good weed management techniques. Pruning eliminates unhealthy or dead material from plants, forms them, and encourages healthy development. Appropriate time and technique are ensured when one is aware of the pruning requirements for various plants. Fertilizing helps plants thrive by restoring nutrients to the soil. Employ organic fertilizers, such as fish emulsion or compost tea, to give your plants a balanced supply of nutrients. In order to prevent pests and diseases, one must keep an eye out for their symptoms and take the necessary precautions. A healthy garden ecosystem can be preserved by implementing integrated pest management (IPM) techniques, which include utilizing physical barriers, promoting beneficial insects, and employing organic insecticides.

Reducing the ecological imprint of the garden and improving environmental health are two benefits of incorporating sustainable techniques into garden layout design. Sustainable gardening places a strong emphasis on promoting biodiversity, conserving water and soil, and using renewable resources. Organic waste, such as leftover food and yard waste, can be composted to produce nutrient-rich compost that can be used to improve soil quality and lessen the need for artificial fertilizers. By adopting companion planting and other natural pest management techniques, including drawing in beneficial insects, chemical pesticide use is reduced, and a healthy environment is maintained. By putting water-saving strategies into practice, such as drip irrigation, rainwater collection, and drought-tolerant landscaping, water resources are preserved, and reliance on municipal water supplies is decreased. Local ecosystems and biodiversity are supported by planting native plants and constructing wildlife habitats like pollinator gardens, bat boxes, and birdhouses.

To sum up, designing a garden layout is a dynamic and thorough process that includes evaluating the site, creating a coherent and useful garden design, choosing suitable plants, preparing the soil, effectively managing water, positioning garden elements, and putting maintenance and sustainable practices into practice. By taking these factors into account, you can design a garden that not only satisfies your functional and aesthetic requirements but also supports a sustainable and healthy environment. Carefully planning your garden will increase its production and enjoyment, enabling you to create a stunning, healthy garden for many years to come.

CHAPTER III

Growing Your Own Vegetables and Herbs

Choosing the right plants for your climate

A key component of successful gardening is selecting plants that are appropriate for your climate. If plants are suited to your area's climate, they will flourish and need less care and resources, creating a garden that is healthier and more sustainable. This section explores the significance of knowing your climate, the many climate zones, the variables influencing plant selection, and the methods for choosing suitable plants for diverse climates, with a focus on designing resilient and fruitful gardens.

The first step in selecting the proper plants is to understand your climate. Climate describes the long-term trends of temperature, precipitation, humidity, and seasonal changes in a particular place. The kinds of plants that thrive in a given area are determined by its distinct climate. Understanding your climate will help you choose plants that are more naturally suited to the area, requiring less frequent and intensive fertilizer, irrigation, and pest control. There are numerous primary classifications of climate, such as tropical, subtropical, temperate, dry, and polar climates. There are unique opportunities and problems for gardeners in each of these climates.

High temperatures and copious amounts of rainfall are year-round features of tropical regions. These areas, which are frequently found close to the equator, are home to a wide variety of plant species and luxuriant flora. Tropical plants often do well in warm, humid areas; they may not survive well in drier or colder regions. When selecting plants for a tropical environment, take into account species that can withstand intense sunlight and high humidity. A few examples are hibiscus, bananas, mangoes, papayas, and many kinds of ferns and palms. Tropical climates with steady warmth and precipitation are ideal for these plants.

Between the tropics and the temperate zones, subtropical climates have hot, humid summers and mild winters. Many different types of plants, including both tropical and temperate species, can be found in these areas. Plants that are able to tolerate the distinct wet and dry seasons that are common in subtropical areas must be chosen. Climates that are subtropical are ideal for citrus trees, bougainvillea, jasmine, and many kinds of cacti and succulents. Not only can these plants withstand summer's heat and humidity, but they can also withstand the rare drought.

Four distinct seasons with changing temperatures and precipitation levels are associated with temperate climates, which encompass much of North America, Europe, and parts of Asia. From chilly winters to balmy summers, these areas provide a wide variety of growing conditions. Think about selecting plants for a temperate region that can withstand temperature swings and seasonal variations. Temperate climates are ideal for deciduous trees like oaks, birches, and maples, as well as perennial flowers like peonies, tulips, and daffodils. These plants are hardy enough to withstand harsh winters and blossom in the spring.

Gardeners face particular difficulties in arid climates because of their high temperatures and little rainfall. These environments, which are frequently deserts or semi-arid regions, call for plants that can endure intense heat and little water. For arid climates, drought-tolerant plants like agaves, succulents, yuccas, and cactus are perfect. Because of their ability to store water in their leaves, stems, or roots, some plants have evolved to withstand protracted dry spells. Furthermore, Mediterranean herbs such as thyme, lavender, and rosemary grow well on poor soils and require little watering.

Extremely low temperatures, brief growing seasons, and sparse vegetation are the hallmarks of polar climates, which are found in areas close to the poles. Plants suitable for cold temperatures and short summers are needed for gardening in arctic regions. Perennials that can withstand cold, such as saxifrages, Arctic poppies, and some grasses, can thrive in polar regions. Hardy vegetables that can withstand the short growing season include root crops, kale, and spinach. In addition to extending the growth season, greenhouses and cold frames can shield plants from inclement weather.

Understanding your climate is not the only thing that affects plant choices; other criteria include soil type, exposure to sunlight, availability of water, and microclimates. Plant health is greatly influenced by the kind of soil since various plants require varied amounts of nutrients and drainage. Plants that are appropriate for your garden's conditions can be selected with the aid of a soil test, which will reveal the pH, texture, and nutrient levels of your soil. For instance, clay soils hold moisture better and are better suited for plants that require moisture, but sandy soils drain fast and are good for drought-tolerant plants.

Sunlight exposure is yet another important consideration when choosing plants. For growth and health, most plants need a certain quantity of sunshine. Partial-shade plants can withstand three to six hours of sunlight every day, but full-sun plants require at least six hours. Less than three hours of direct sunshine are sufficient for shade-tolerant plants to thrive. Plant health and productivity can be guaranteed by keeping an eye on your garden's lighting patterns and matching plants to the right amount of light.

Plant growth depends on the availability of water, and the needs of individual plants vary. Since they don't require much watering, drought-tolerant plants are perfect for areas with scarce water supplies or erratic rainfall. On the other hand, plants that prefer moisture need regular irrigation or rainfall and are more suited to areas with copious amounts of rainfall. Water waste can be minimized, and water-efficient zones can be created by grouping plants with comparable water requirements together.

Plant selection may also be impacted by the microclimates in your garden. Microclimates are small regions with particular environmental characteristics, like differences in humidity, temperature, or wind exposure. A gloomy corner may stay cooler and more humid, making it ideal for plants that can withstand shade, but a south-facing wall can generate a warm, protected microclimate that favors plants that thrive in heat. You may cultivate a wider variety of plants and design more resilient and varied garden areas by identifying and making use of microclimates in your yard.

It is important to learn about the unique requirements and traits of every plant species in order to choose plants that are suitable for different climates. The best plants for your climate can be determined by consulting local gardening resources, including botanical gardens, extension agencies, and knowledgeable gardeners. Plant hardiness zones, or the range of climatic conditions a plant can withstand, are another resource for knowledge about plants: gardening books, internet databases, and plant nurseries.

Gardeners can benefit greatly from the United States Department of Agriculture's (USDA) establishment of plant hardiness zones. Based on the average annual minimum winter temperature, these zones are split into 13 groups, each of which represents a range of 10 degrees Fahrenheit. It is easier to choose plants that can withstand the winter temperatures in your area if you are aware of your hardiness zone. A plant designated for hardiness zone 5, for instance, is suited for areas with cold winters since it can tolerate minimum temperatures as low as -20°F. A plant designated for hardiness zone 10, on the other hand, is perfect for warmer locations because it can withstand minimum temperatures as low as 30°F.

In addition to hardiness zones, other aspects, including insect resistance, humidity preferences, and heat tolerance, should be taken into account while choosing plants. For areas where summers are hot, plants must be tolerant of heat since some species cannot withstand high temperatures. Plants differ in their preferred levels of humidity; some do well in damp environments, while others do well in dry air. Additionally, pest resistance is crucial since some plants are more vulnerable to diseases and pests that are prevalent in particular climates. Choosing plants that are naturally resistant to nearby pests can help create a healthier garden environment and lessen the need for chemical treatments.

Another tactic that can improve plant choice and garden health is companion planting. Growing specific plants next to each other for mutual benefit—such as enhanced growth, pollination, or insect control—is known as companion planting. Growing marigolds next to tomatoes, for instance, can help keep nematodes away, and growing basil next to peppers can improve flavor and keep pests away. By increasing diversity and resilience through companion planting, you may lessen your reliance on outside resources and foster ecological balance in your garden.

Making a garden plan that takes into account the plants' growth habits and spacing needs is just as important as choosing the appropriate plants. Plants with the right spacing will be able to grow in sufficient space, get enough sunlight, and get nutrients and water. Competition for resources, a rise in pest and disease pressure, and a decline in plant productivity and health can all result from overcrowding. A balanced and flourishing garden can be achieved by investigating the mature size of each plant and designing your garden layout appropriately.

Complementing plant selection and enhancing the general health and sustainability of your garden are sustainable gardening techniques like mulching, composting, and integrated pest management. Adding organic materials, like wood chips, straw, or leaves, to the soil helps control temperature, retain moisture, and keep weeds at bay. The process of composting organic waste, which includes yard trash and kitchen scraps, produces nutrient-rich compost that can enhance the fertility and structure of the soil. In order to control pests and illnesses, minimize the need for chemical pesticides, and foster a healthy garden ecology, integrated pest management, or IPM, combines mechanical, biological, and cultural approaches.

It also takes responding to shifting climatic conditions to create a robust and fruitful garden. Because climate change can affect temperature patterns, precipitation levels, and the frequency of extreme weather events, gardeners face considerable problems. It will take adaptability and planning in plant selection and garden maintenance to keep up with these changes. A more robust garden that can tolerate shifting conditions can be achieved by using a wide variety of plants with different climate tolerances. Furthermore, putting water-saving strategies into practice, such as drip irrigation and rainwater collection, can lessen the effects of drought and water scarcity.

Finally, selecting plants that are appropriate for your climate is a complex process that includes learning about your local climate, taking into account elements like soil composition, exposure to sunlight, water availability, and microclimates, and utilizing techniques like companion planting, plant hardiness zone research, and sustainable gardening methods. A resilient, productive, and sustainable garden can be created by choosing plants that are appropriate for your environment and designing the layout of the garden to fit their requirements. Carefully choosing plants and maintaining your garden will improve its aesthetics and usefulness while encouraging sustainability and environmental care. For your garden to remain successful and healthy over the long term, it will be more crucial than ever to select plants that are resilient and adaptable to changing climatic circumstances.

Starting from seeds vs. transplants

As a fun and fulfilling endeavor, planting a garden gives you the chance to cultivate your own fruits, vegetables, flowers, herbs, and more. Whether to use transplants or grow plants from seeds is one of the most important decisions gardeners have to make. Both approaches have benefits and drawbacks, and the decision is influenced by a number of variables, including the gardener's background, the resources at hand, the climate, and particular gardening objectives. In order to assist gardeners in making wise selections, this section will compare and contrast starting from seeds with using transplants, looking at the advantages, disadvantages, and best practices of each method.

Planting seeds directly in the ground or growing them in pots indoors before moving the seedlings to their final site is known as "starting from seeds." With this technique, gardeners can select from a large range of plant types and cultivars, frequently having more control over the whole growth process. The wide variety of plants that can be grown is one of the main advantages of starting from seeds. Thousands of seed kinds are available to gardeners, including hybrids, heirlooms, and unusual or rare plants that might not be available as transplants. Because of this diversity, gardeners can more fully customize their space and cultivate plants that are suited to their unique preferences, tastes, and growth environments.

The affordability of beginning from seeds is another benefit. Gardeners can save money by cultivating a variety of plants or planting huge areas by using seeds instead of transplants, which are typically much more expensive. Furthermore, seeds can be preserved for later use, offering a long-term and sustainable gardening solution. Beginning with seeds also gives you the gratification of raising plants from seed to maturity, giving you a stronger bond with gardening and a sense of achievement.

Starting from seeds does, however, come with a number of difficulties. The requirement for cautious management and close attention to detail throughout the germination and early growth stages is one of the primary challenges. For seeds to germinate effectively, certain circumstances must be met, such as the ideal levels of light, moisture, and temperature. Certain seeds need certain conditions to break dormancy and encourage germination, such as stratification (cold exposure) or scarification (seed coat breaking). Particularly for indoor seed beginnings, maintaining these conditions might take time and may call for extra tools like humidity domes, grow lamps, or heat mats.

Moreover, developing plants from seeds frequently requires a longer growing season before they reach maturity. As seeds may take several weeks to germinate and extra weeks or months to grow into healthy, productive plants, gardeners must exercise patience and diligence. It can be difficult to follow this longer schedule, particularly in areas with short growing seasons. To get around this, gardeners might have to sow seeds indoors far in advance of the last frost date in order to guarantee that seedlings are prepared for transplanting when the weather is suitable outside.

Using transplants, on the other hand, entails buying or cultivating young plants that are already established and prepared for garden planting. There are a number of advantages to this approach, such as time savings, convenience, and higher success rates. The head start that transplants offer is among their main benefits. Gardeners may plant transplants straight into the garden with confidence because they have already successfully completed the crucial germination and early growth stages. Beginners or those with little time or resources to handle the difficulties of seed beginning may especially benefit from this.

Gardeners can also extend the growing season and reap earlier harvests with the help of transplants. Gardeners can enjoy their crops earlier in the season and take advantage of ideal growth circumstances by starting with young plants that are already a few weeks old. For warm-season crops like tomatoes, peppers, and eggplants, which need a long growing season and would not mature in time if planted from seeds in areas with short growing seasons, this is especially crucial.

Transplanting also makes it possible to use garden areas more effectively. Transplants may be placed appropriately from the start because they are already established, which minimizes the requirement for thinning and wasted space. Increased yields and more productive gardens can result from this, especially in compact or urban environments where space is at a premium.

Despite these advantages, employing transplants has disadvantages as well. The restricted variety of plant kinds accessible for transplantation is one of the primary drawbacks. While there are many common and popular plants available at nurseries and garden centers, the selection is frequently more limited than the wide variety of seed options. It can be difficult for gardeners looking for rare or specialized plants to acquire transplants that fit their demands.

Additionally, transplants typically cost more than seeds. Buying young plants may get expensive, especially if you're cultivating a lot of different plant species in a large garden. For individuals on a tight budget or wishing to get the most out of their gardening investment, this cost may be very important.

The possibility of transplant shock is another possible problem. Young plants may become stressed when transplanted from containers into the garden because of environmental, handling, and soil changes. If this stress is not adequately handled, the plant may experience wilting, slower growth, or even die. Hardening off or acclimating transplants gradually to their new surroundings is a crucial step gardeners must take to reduce transplant shock. In order to assist the plants adjust and become less stressed, this entails exposing them to outdoor circumstances gradually over a few days.

Choosing between starting from seeds or utilizing transplants depends on a number of elements, such as the gardener's goals, experience, resources, and environment. Each method has advantages and disadvantages of its own. The best course of action is frequently to combine the two approaches, which enables gardeners to benefit from each method's advantages while minimizing its disadvantages.

For instance, in order to give slow-growing or long-season crops like broccoli, tomatoes, and peppers enough time to mature and yield a harvest, gardeners may decide to start these crops from seeds indoors. In order to jump-start the garden and get earlier yields, people may also buy transplants of cool-season or fast-growing crops like lettuce, spinach, and herbs. With this hybrid technique, time and resources are optimized while garden flexibility and efficiency are increased.

It is imperative to adhere to recommended practices when starting from seeds in order to guarantee good seedling development and successful germination. Start by choosing premium seeds from reliable vendors, and follow the suggested planting dates and guidelines on the seed packet. To ensure the proper ratio of nutrients and drainage and to ward off illness, use a sterile seed-starting mix. Maintain a constant moisture content in the soil without becoming too wet, and plant seeds at the proper depth and spacing.

Enough light must be provided for indoor seed beginning. For the majority of seeds to develop robust and healthy, 12 to 16 hours of light per day are needed. Use grow lights to augment natural light if it's not enough to give seedlings the right amount of light. To encourage germination and growth, maintain the right temperature and humidity levels; if necessary, use heat mats or humidity domes.

Feed seedlings frequently with a diluted, balanced fertilizer as they grow to aid in their development. Following the correct hardening-off processes to accustom them to outside circumstances, seedlings can be moved into larger containers or straight into the garden once they have several sets of true leaves and are large enough to handle.

It is important to choose robust, healthy plants from reliable nurseries or garden centers if you are one of those people who would rather use transplants. Seek transplants that have robust root systems, colorful foliage, and no evidence of insect or disease damage. Steer clear of plants that are lanky, wilting, or root-bound since they can find it difficult to take root and flourish in your yard.

To increase drainage and fertility before planting transplants, prepare the garden bed by adding organic matter, such as compost, to the soil. Make sure not to injure the roots when removing the plant from its container by carefully digging holes that are only slightly bigger than the transplant's root ball. After planting the transplant in the hole at the same depth as it was growing in the container, gently compact the dirt around the roots with your hands. After planting, give the transplant plenty of water to aid in settling in and lessen transplant shock.

In conclusion, there are clear benefits and drawbacks to both beginning from seeds and employing transplants. Although seeds come with many plant options, are less expensive, and provide you the satisfaction of starting a plant from scratch, they also come with some caution and waiting time. Although they save time, are more convenient, and give you a head start on the growing season, transplants are more expensive and have a smaller variety range. Gardeners can customize their strategy to fit their unique conditions and requirements by knowing the advantages and disadvantages of each method. By combining the two approaches, one can maximize garden productivity and enjoyment while getting the best of both worlds. The decision to start from seeds or use transplants ultimately comes down to the objectives, means, and tastes of the gardener; either way can result in a fruitful and satisfying gardening endeavor.

Seasonal planting guide

By coordinating agricultural operations with the varying seasons, seasonal planting makes sure that crops are planted, cultivated, and collected at the most advantageous times. The complexities of seasonal planting are explored in this guide, which looks at the interactions between various soil types, climates, and crops throughout the growing season. Planting schedules that are in line with the environment's natural cycles help farmers and gardeners increase yields, lower pest and disease populations and build a more resilient agricultural system.

Finding out what makes each season unique is the first step toward comprehending seasonal planting. The year is normally split into four seasons in temperate regions: spring, summer, autumn, and winter. Plant development and health are influenced by the distinct conditions that are presented by each season. Spring is the best season to plant seeds and transplants because of the longer days and higher temperatures. Summertime promotes quick development and fruiting because of its long days and high temperatures. Fall offers shorter days and colder temperatures, which help late-season crops mature and grow their roots. Many plants become dormant during the winter months, which are frequently characterized by frost and below-freezing temperatures, while some resilient crops can be cultivated during this period.

A wide range of crops can be planted in the spring because the earth starts to warm up, and there is less chance of frost. Cool-season veggies like lettuce, spinach, peas, and radishes grow well in the mild early spring temps, making this the perfect time of year to plant them. Warm-season vegetables like tomatoes, peppers, cucumbers, and beans can be directly seeded into the garden or moved as the season goes on and the temperature rises. Before planting, the soil must be properly amended with enough organic matter and nutrients to promote strong development. Mulching can improve moisture retention, control weed growth, and make the soil more plant-friendly for seedlings.

The growing season peaks in the summer, when photosynthesis and growth are stimulated by long days and abundant sunlight. Warm-season crops are at their peak during this time, yielding bountiful harvests of fruits, vegetables, and herbs. Summertime's heat and dryness, however, can bring with it difficulties, like higher water needs and the risk of heat exhaustion. Plants require regular irrigation to stay healthy, and techniques like soaker hoses and drip irrigation can minimize evaporation and provide moisture directly to the roots. Shade cloths or row covers can shield delicate plants from direct sunlight and extreme heat. Furthermore, summer is a crucial time for managing pests and diseases because of the favorable conditions that warm temperatures and high humidity can generate for the growth of pathogens and insects. A healthy garden ecosystem can be preserved with the use of integrated pest management (IPM) techniques such as crop rotation, beneficial insects, and organic insecticides.

The change from summer heat to winter chill arrives in the fall. Cool-season crops, which can withstand minor frosts and flourish in the milder weather, are best planted during this time of year. Autumn planting works well for leafy greens like kale, chard, and arugula, as well as root crops like carrots, beets, and turnips. Because of the lower temperatures and fewer days, plants are able to establish strong root systems and accumulate sugars, which improve the flavor of many foods. To increase soil richness and structure, fall is a good time to plant cover crops, also referred to as green manures. These crops, like rye, vetch, and clover, prevent soil erosion, control weed growth, and fix nitrogen for the benefit of the next growing season.

For farmers and gardeners, winter offers a special combination of opportunities and problems. Hardy crops like kale, Brussels sprouts, and leeks can grow and supply fresh vegetables all season long in areas with moderate winters. Using cold frames, greenhouses, or hoop houses to prolong the growing season and shield plants from inclement weather is a common practice in winter gardening in colder climates. Greens and other cold-tolerant crops can be grown in these buildings because they produce a microclimate that can increase temperatures by several degrees. Winter is also a great time to plan for the forthcoming growing season and prepare the soil. By measuring pH and nutrient levels, soil tests can help guide amendments to enhance soil health. Wintertime is a great time to continue mulching and composting, which will supply important organic matter for spring planting.

It is crucial to take into account the unique requirements of various crops in addition to comprehending the seasonal dynamics of planting. Every plant has specific needs for moisture, light, temperature, and nutrients, all of which affect the best time to plant and grow. For instance, severe heat can harm cool-season crops like broccoli and cauliflower, which prefer temperatures between 60 and 70°F (15 and 21°C). Warm-season crops, on the other hand, like okra and melons, need a long growing season to achieve maturity and flourish in temperatures above 70°F (21°C). A more effective and productive garden can be created by gardeners by aligning crop requirements with seasonal circumstances.

Another essential component of good seasonal planting is healthy soil. Strong plant growth is supported by healthy soil, which gives vital nutrients, water, and air. Techniques like crop rotation, cover crops, and the addition of organic matter can improve the fertility of the soil. Crop rotation is the practice of growing various crop families in different locations each year to balance nutrient consumption and prevent the accumulation of pests and illnesses. As was previously noted, cover crops increase microbial activity, contribute organic matter, and improve soil structure. Enhancing soil health through composting, mulching, and applying organic fertilizers can also lead to a flourishing garden environment.

Considering the environment and microclimate is essential when organizing a seasonal planting program. The suitability of various crops and the timing of planting are determined by the local climate, which includes typical temperatures, rainfall patterns, and dates of frost. Small regions with unique climates, known as microclimates, are influenced by a variety of elements, including urban settings, water bodies, slopes, and elevation. Gardeners may choose the best crops for each location and plant dates by being aware of these variations. In contrast, a low-lying location can be vulnerable to frost and need frost-tolerant crops or protection measures. For instance, a south-facing slope might warm up quicker in the spring and be perfect for early planting.

Because changing seasons bring with them varied water requirements and obstacles, water management is an essential component of seasonal planting. For young plants to establish themselves and for seeds to germinate in the spring, there must be constant wetness. Summertime temperatures and evaporation rates often necessitate more frequent watering. Effective water management techniques include mulching, drip irrigation, and crop selection that is resistant to drought. Autumn usually provides more rain, which lessens the need for additional watering, but in order to avoid waterlogging, appropriate drainage must be ensured. Water requirements are lower in the winter, but it's still important to keep an eye on soil moisture levels and give water when it's dry outside, particularly for crops and perennials grown in covered areas.

Seasonal planting also requires careful attention to pest and disease management. Seasonal variations in pest pressures necessitate a combination of preventive and reactive management techniques for optimal results. Aphids, beetles, and caterpillars are examples of insect pests that are most prevalent in the spring and summer and can be controlled by crop rotation, companion planting, and the use of physical barriers. To organically manage pest populations, beneficial insects can be introduced, such as ladybugs and lacewings. Disease management includes choosing disease-resistant cultivars and, when needed, applying organic fungicides in addition to preserving plant health through appropriate spacing, watering, and sanitation techniques.

In summary, seasonal planting is a dynamic and intricate technique that necessitates knowledge of the environment's natural cycles as well as the unique requirements of various crops. Planting schedules that are in line with seasonal conditions help farmers and gardeners minimize pest and disease pressures, maximize growth, and build a more sustainable agricultural system. For anyone wishing to improve their farming or gardening techniques, this guide offers a thorough explanation of the concepts and procedures of seasonal planting. Embracing the rhythms of the seasons can result in more abundant harvests and a closer bond with nature, regardless of expertise level in farming.

Tips for maximizing yield in small spaces

In the fields of urban gardening and small-scale farming, maximizing productivity in limited areas is a topic that is becoming more and more significant. Food security and sustainability depend on our ability to cultivate a large number of crops in small spaces, given our expanding population and a finite amount of available land. This in-depth section will examine a range of tactics and methods to increase food yields in tiny areas, including effective use of available space, vertical gardening, managing soil, choosing crops, intense planting techniques, container gardening, and cutting-edge technologies.

Effective use of available space is a key factor in optimizing yield in compact settings. Conventional gardening techniques frequently emphasize horizontal planting, but constrained areas need for a more calculated strategy. For example, raised beds can work incredibly well in tiny gardens. Because these beds are usually 4 feet wide, soil compaction is reduced, and access from both sides is made simple. In addition to extending the growing season, improving drainage, and warming up more quickly in the spring, raised beds also increase yields. Gardeners may maximize plant spacing and design a more productive garden layout by arranging crops in raised beds.

The use of vertical gardening is another effective method for increasing productivity in compact areas. Gardeners can greatly expand their growth area by growing plants upward instead of outward. Climbing plants like beans, peas, cucumbers, and tomatoes can be supported by trellises, arbors, and vertical containers. This technique reduces the danger of fungal diseases by improving air circulation around the plants and saving space. Additionally, because plants are positioned to absorb more light, vertical gardening can improve exposure to sunshine. To maximize vertical space for non-climbing plants, wall-mounted planters, hanging baskets, and stacking containers can be used.

Optimizing production requires careful attention to soil management, particularly in small areas with limited soil resources. Good, nutrient-rich soil is the starting point for successful gardening. Gardeners should concentrate on adding organic matter, such as compost, aged manure, and leaf mold, to the soil in order to improve its fertility and structure. These amendments improve the texture of the soil, boost its ability to hold water, and give plants the vital minerals they need. Conducting routine soil testing can aid in determining nutrient deficits and provide direction for applying organic fertilizers. Mulching is another useful technique since it retains moisture, keeps weeds at bay, and enriches the soil with organic matter as it breaks down.

Choosing the right crops is essential to maximizing yields in tiny gardens. Selecting space-efficient, high-yielding cultivars can have a big impact. For example, determinate tomato types are better suited for limited spaces since they reach a specific height and set fruit all at once, whereas indeterminate kinds grow and produce fruit all season long. Vegetables like beans, peas, and zucchini that grow as dwarf or bush kinds work well in small spaces as well. To further increase productivity, choosing crops with shorter growing seasons can enable repeated harvests in a single season. Another effective tactic is companion planting, which involves grouping plants with similar growth patterns and nutritional requirements to make the most of available space and resources.

In tiny areas, intensive planting techniques like square-foot gardening and succession planting can greatly increase yields. Using square-foot gardening, the garden is divided into portions that are each square foot in size, and crops are planted densely within each area. This technique reduces wasted space and makes the most of the available space. Additionally, it makes pest control and crop rotation easier. Crops are sown successively in succession planting in order to guarantee a consistent harvest all through the growing season. Gardeners can replant the area with mid-season crops like beans or carrots after harvesting early-season crops like lettuce or radishes. By maximizing the use of time and space, this technique raises yields overall.

For people with extremely little space, such as those with balconies, patios, or tiny urban backyards, container gardening is a great solution. Numerous types of vegetables, herbs, and even tiny fruit trees can be grown in containers. Choosing the appropriate size and kind of container for each plant is essential to successful container gardening. Greater room for root development and less frequent watering are provided by larger pots. To avoid waterlogging, it's also critical to use premium potting soil and make sure drainage is adequate. Maintaining constant moisture levels is essential for the best possible plant growth and productivity, and drip irrigation systems and self-watering containers can help.

New methods and technology are always being developed to assist gardeners in increasing yields in constrained areas. For instance, hydroponics is a soilless growth technique that involves growing plants in nutrient-rich water solutions. This method is perfect for limited spaces since it can be put up vertically and permits high-density planting. Compared to conventional soil-based gardening, hydroponics systems can offer far better yields and can be customized for a variety of crops. Another cutting-edge technique is aquaponics, which combines hydroponics and aquaculture (fish farming). In this arrangement, the plants help filter and clean the water for the fish, and the waste from the fish gives the plants nutrition. A symbiotic environment that maximizes production and resource use is produced by aquaponics.

Another crucial element in optimizing yields is light management, especially in cities where buildings and other structures can shade garden sections. Plants can receive more light when they are exposed to reflective surfaces like white walls or mulch. Grow lights can also be used to supplement natural sunshine, particularly in enclosed or darkened settings. Energy-efficient LED grow lights can be tailored to give the precise light spectrum required for a plant's stage of growth. Plant growth and productivity are fueled by photosynthesis, which depends on sufficient light for photosynthesis to occur.

Water management in small-space gardening is similarly crucial. By delivering water directly to the plant roots, efficient irrigation systems—like soaker hoses or drip irrigation—minimize water wastage and the chance of fungal diseases. By offering a sustainable water source, rainwater harvesting devices can lessen dependency on municipal water supplies. Mulching helps hold onto soil moisture, which lessens the need for regular irrigation. Self-watering systems in container gardening can keep moisture levels constant, avoiding the stress brought on by over- or under-watering.

Controlling pests and diseases is essential to keeping plants healthy and increasing harvests. It's critical to often check plants in tiny places for indications of pests and illnesses. Biological, mechanical, and cultural treatments combined in integrated pest management (IPM) strategies can effectively control insect populations. Aphid populations can be controlled, for instance, by introducing beneficial insects like ladybugs and lacewings. Plants can be shielded against insect pests by using nets or row coverings. Even in tiny areas, crop rotation can help stop the spread of illnesses that are borne on the soil. It's also critical to keep your garden clean and tidy by getting rid of any unhealthy plants and disinfecting your tools. This will stop infections from spreading.

In tiny gardens, companion planting is an excellent way to increase harvests. Some plants can benefit their neighbors' development and health by deterring pests, drawing helpful insects, or offering support and shade. For instance, marigolds can ward off nematodes, while basil can ward against tomato hornworms. Beans and peas are examples of legumes that fix nitrogen in the soil, which helps nearby plants that require more nitrogen. Gardeners can build a more resilient and productive ecosystem in their gardens by carefully choosing and arranging companion plants.

Another crucial technique for boosting yields and preserving soil health is crop rotation. Rotating crops every season can help balance nutrient consumption and minimize the accumulation of pests and illnesses, especially in tiny plots. To maximize soil fertility, for instance, nitrogen-fixing legumes should be planted after heavy feeders like corn or tomatoes. You can disrupt the cycles of pests and diseases by rotating root crops with leafy greens. Over time, increasing garden output can be achieved by planning efficient crop rotations and keeping thorough records of planting dates, locations, and crop performance.

Gardeners that use season extension techniques can grow crops much beyond the normal growing season, which can result in significantly higher yields. Cool-season crops can be cultivated in winter, and warm-season crops can be started early in the spring thanks to the protection from frost that cold frames, hoop houses, and greenhouses offer. Plants can be shielded from early frosts and have their harvest season extended with row covers and cloches. Gardeners can maximize their modest spaces and reap several harvests by prolonging the growing season.

In tiny gardens, pruning and training plants can also help to maximize output. In order to focus energy on producing fruit, pruning entails cutting off any extra growth, such as suckers on tomato plants or side shoots on pepper plants. Getting plants to grow vertically on trellises or other supports can enhance space usage, air circulation, and sunlight exposure. For instance, teaching melons and cucumbers to climb trellises might clear space on the ground for other crops. A more productive and well-organized garden can be achieved by routinely tracking and modifying plant growth.

Perennial plants can add long-term productivity to a small garden and lessen the need for seasonal replanting. Berries and fruit trees, as well as perennial veggies like rhubarb and asparagus, can supply for many years. These plants can be incorporated into vertical structures or garden borders and often require less upkeep. Gardeners may construct a more sustainable and diverse garden that yields a consistent source of food all year round by combining annual and perennial crops.

In small locations, maximizing yield also necessitates meticulous organization and planning. To make the most of available space and resources, a thorough garden plan that includes planting dates, crop locations, and succession planting plans should be created. Recording plant kinds, growth characteristics, and insect problems helps with garden management by providing information for future planting selections. Gardeners may concentrate on plant care and upkeep by saving time and effort by efficiently organizing their tools and supplies.

Another strategy for increasing yields in cities with little personal space is community gardening. More growing space and resources may be accessible through community-supported agriculture (CSA) initiatives, rooftop gardens, and shared garden plots. Working together with other gardeners can also make it easier to share tools, seeds, and information, which will increase small-space gardening's overall productivity and sustainability. Community gardens frequently function as centers of teaching, supporting programs related to urban agriculture and food security.

To sum up, optimizing productivity in compact areas necessitates a diverse strategy that includes effective use of available space, vertical gardening, crop selection, soil management, intense planting techniques, container gardening, and cutting-edge technologies. Even in the most constrained spaces, gardeners may develop extremely productive and sustainable gardens by implementing these principles as well as ongoing learning and adaptation. The ability to produce an abundance of crops in compact places will become more and more important for maintaining food security and encouraging sustainable living as urban populations continue to rise and land availability declines. Small-space gardeners can produce amazing results and help create a more resilient and sustainable food system by using creative tactics, resource management, and careful planning.

CHAPTER IV

Fruits, Berries, and Nut Trees

Selecting fruit trees and bushes for your backyard

Choosing fruit trees and shrubs for your backyard can turn your outdoor area into a beautiful and fruitful garden. It's a gratifying project. This section examines a number of factors that should be taken into account when selecting fruit trees and shrubs, such as soil needs, climate, variety selection, area planning, planting methods, and upkeep procedures. Gardeners can create a successful backyard orchard by having a thorough awareness of these variables and making wise judgments.

Examining your local climate is the first step in choosing fruit trees and bushes for your backyard. Specific climate needs, such as temperature ranges, chilling hours, and frost tolerance, apply to different fruit species and cultivars. The total number of hours below 45°F/7°C that a fruit tree needs to break dormancy and provide blossoms and fruit is referred to as its "chilling hours." A key to effective production is knowing the environment of your area and combining the right fruit kinds with it.

Many fruit trees, including those that bear apples, pears, plums, cherries, and peaches, flourish in temperate areas because of the unique seasonal variations that supply the required number of chilling hours. For example, depending on the variety, apples can need 500–1,000 hours to chill. Gardeners may choose low-chill kinds or fruit trees that have adapted to such conditions naturally in warmer locations with mild winters and a few hours of chilling. Due to their lower chilling hour requirements and tolerance for higher temperatures, citrus trees—which include orange, lemon, and lime—as well as figs and pomegranates—make ideal selections for warm areas.

Another important consideration while choosing and cultivating fruit trees and shrubs is the quality of the soil. In general, fruit plants want soil that is healthy, well-drained, and has a balanced pH of 6.0 to 7.0. Before planting, a soil test can yield important details regarding the pH, nitrogen levels, and organic matter content of the soil. Gardeners can establish ideal growing conditions by amending the soil with compost, lime, sulfur, or other organic fertilizers based on the results. In regions with a lot of clay or poorly drained soil, raised beds or mounds can help with drainage.

To make sure that fruit trees and bushes have enough space to grow and get enough sunshine, space planning is crucial. The majority of fruit trees need full sun, or six to eight hours of direct sunlight each day. Fruit trees should be chosen with their mature size and spacing needs in mind. Smaller backyards can accommodate dwarf and semi-dwarf fruit trees, which are more compact than standard fruit trees, which can grow to a height of 20 to 30 feet and require a lot of room between them. Dwarf trees are perfect for small spaces or even container gardening because they may be planted closer together and normally reach a height of 8 to 10 feet.

Selecting the best fruit trees and bushes requires taking into account a number of variables, including harvest schedules, disease resistance, flavor preferences, and pollination requirements. Since flavor and texture are subjective, it's critical to choose kinds that appeal to you. Another important factor to take into account is disease resistance since certain cultivars are bred to resist common diseases and pests, which lowers the need for chemical treatments. For example, peach types like 'Redhaven' and 'Elberta' have good resistance to bacterial spot, while some apple varieties, including 'Enterprise' and 'Liberty,' are noted for their resistance to apple scab and fire blight.

Because some fruit trees can produce fruit on their own while others need cross-pollination from a different type in order to set fruit, pollination is an important consideration when choosing fruit trees. Self-pollinating trees are ideal for small gardens since they can bear fruit on their own. Examples of these trees include most peaches, nectarines, and sour cherries. But cross-pollination is necessary for many apples, pears, plums, and sweet cherries. To guarantee effective pollination and fruit set, plant at least two distinct kinds of the same fruit species that bloom at the same time. To increase yields, use pollination charts and select appropriate types.

Another crucial factor to take into account is when to harvest fruits. Throughout the growing season, gardeners can have a continual harvest by choosing a combination of early, mid, and late-season cultivars. 'Lodi' and 'Gala' are early-season apple types that mature in late summer; 'Honeycrisp' and 'Jonathan' are mid-season kinds that ripen in early fall; and 'Granny Smith' and 'Fuji' are late-season varieties that ripen in late fall. In addition to ensuring a consistent supply of fresh fruit, this spaced harvest lowers the possibility of fruit waste and excess.

It takes careful planning and the right methods to plant fruit trees and shrubs in order to ensure their good establishment and growth. Before the buds open, the dormant season, which is usually late winter or early spring, is the ideal time to grow fruit trees. Because of this timing, the roots can take hold before the arrival of warm weather. During this time, bare-root trees—which are marketed without soil surrounding their roots—are frequently planted. Although trees grown in containers can be planted any time of year, it is best to do so in the early spring or fall to protect the plants from the harsh winter weather.

Dig a hole that is exactly as deep and twice as wide as the root ball when planting a fruit tree. This gives the roots plenty of room to grow and spread. Place the tree in the hole so that the graft union, or the enlarged point where the scion and rootstock converge, is higher than the soil line. To remove any air pockets, carefully compact the native soil surrounding the roots as you backfill the hole with it. After planting, give the tree plenty of water to let the soil settle, and the roots get moist. Mulch applied around the tree's base helps control soil temperature, hold on to moisture, and keep weeds at bay.

Fruit trees and shrubs require regular pruning in order to stay healthy and productive. Frequent trimming improves air circulation, shapes the tree, and gets rid of unhealthy or dead limbs. Depending on the kind of fruit tree, different pruning schedules and techniques apply. For example, pruning is best done in the summer to control the size and shape of peach and cherry trees and during the dormant season on apple and pear trees to promote fresh growth and fruit production. The structure and vigor of the tree are managed using pruning procedures, including heading cuts, which shorten branches, and thinning cuts, which remove entire branches.

Fertilization is essential to supply the nutrients required for the growth and fruit production of fruit trees and bushes. It is customary to apply a balanced fertilizer that has the right amounts of potassium (K), phosphorus (P), and nitrogen (N). Potassium improves fruit quality and overall plant health, phosphorus helps root development and flowering, and nitrogen encourages the growth of leaves. To increase soil fertility, organic fertilizers like compost, manure, and bone meal can be applied. In order to prevent overfertilization, which can result in excessive vegetative growth and decreased fruiting, it is crucial to adhere to approved treatment rates and timing.

Another crucial component of caring for fruit trees is irrigation. For newly planted trees to develop robust root systems, frequent irrigation is necessary. It is more efficient to water the soil deeply—a few inches down—than to irrigate it shallowly on a regular basis. Fruit trees require 1 to 2 inches of water each week once they are established, either from additional irrigation or rainfall. Water loss from evaporation can be reduced by using soaker hoses or drip irrigation systems, which effectively supply water to the root zone.

A vital part of keeping fruit trees and bushes healthy is managing pests and diseases. Aphids, scale insects, codling moths, and spider mites are frequent pests; powdery mildew, apple scab, fire blight, and brown rot are prevalent diseases. Biological, chemical, and cultural controls are combined in integrated pest management (IPM) techniques, which effectively manage diseases and pests. Pest and disease pressure can be decreased by implementing cultural measures like crop rotation, good sanitation, and pruning. Beneficial insects and microbial treatments are examples of biological controls that can assist in naturally regulating pest populations. Use chemical controls, such as synthetic and organic insecticides, only as a last option and in accordance with label directions.

Timely harvesting of fruit is crucial to obtaining optimal flavor and quality. Distinct fruit varieties and types undergo distinct ripening processes. Apples, for instance, are usually picked when they are firm, have a slightly changed color, and have the right amount of sweetness and acidity. The optimal time to harvest peaches is when they smell good and give slightly light pressure. Strawberries, raspberries, and blueberries are examples of berries that should be completely colored and capable of being gently pulled off the vine. Harvesting fruit should be done carefully to prevent bruising and damage, which can reduce its shelf life.

The last stage of planting fruit trees and bushes is to preserve and enjoy the produce. You can eat fresh fruit right away or save it for later. Many fruits can have their shelf lives extended under the right storage circumstances, which include high humidity and cold temperatures. For several months, apples, for instance, can be kept in a cold, dark spot. In addition, fruits can be preserved via canning, freezing, drying, and creating jams and jellies. Gardeners can continue to savor the aromas of their produce long after the growing season has concluded, thanks to these preservation techniques.

In conclusion, careful consideration of temperature, soil, area, and variety selection is necessary when choosing fruit trees and shrubs for your garden. Gardeners can establish a fruitful and delightful backyard orchard by comprehending these elements and putting appropriate planting and upkeep techniques into practice. Selecting, planting, and maintaining fruit trees and bushes is a fulfilling experience that yields fresh fruit from your own garden and improves the garden's aesthetics and biodiversity. Anybody can successfully cultivate a variety of fruit trees and shrubs in their backyard, helping to promote a self-sufficient and sustainable lifestyle with careful planning and hard work.

Planting and caring for perennial plants

Perennial plant planting and maintenance is an essential part of gardening that provides long-lasting aesthetic appeal, ecological advantages, and sustainability. Unlike annuals, which need to be replanted every year, perennials, which live for more than two years, return season after season with little care. The entire realm of perennial gardening is covered in this section, including the selection process, soil preparation, planting methods, upkeep procedures, control of pests and diseases, and the aesthetic and ecological advantages they offer to gardens and landscapes.

The first step to designing a great perennial garden is choosing the appropriate perennial plants. Understanding your garden's unique growing conditions—such as its climate, kind of soil, exposure to light, and moisture content—is necessary for this procedure. Since every perennial has different requirements and tolerances, it's critical to match plants with the ideal environment to ensure their survival and growth. Certain plants, like lavender and sedum, appreciate sunny, well-drained spots, but others, like hostas and ferns, do well in moist, shaded situations. Perennials should be carefully considered for their mature size to make sure they blend in with the garden and don't crowd out other plants.

When choosing perennial plants, climate is an important factor. One useful resource that gardeners can use to identify which perennials are appropriate for their area is the USDA Hardiness Zone map. Based on average annual minimum winter temperatures, this map splits North

America into zones, with a 10°F variation between each zone. To make sure they can survive the harsh winters in their area, gardeners should select perennials that are rated for their particular zone. For instance, a perennial that does well in Zone 5 might not withstand Zone 3's harsher winters or Zone 8's milder climate.

For perennial plants to thrive, the soil must be prepared. Sturdy, well-drained soil is the basis for robust root systems and fast development. It is crucial to test the soil to find out its pH and nutrient levels before planting. Most perennials like a pH range of 6.0 to 7.0, which is slightly acidic to neutral. The pH of the soil can be changed by adding lime to increase it or sulfur to drop it if it is too high or too low. Furthermore, adding organic matter to the soil—such as aged manure or compost—improves drainage, fertility, and soil structure. Adding organic matter to the soil also encourages good microbial activity, which improves plant health and nutrient availability.

Establishing perennial plants requires careful planting methods. Spring or fall, when temperatures are milder, and there is less chance of excessive heat or cold, are the ideal seasons to plant perennials. Because of this timing, perennials can establish their root systems before summer or winter damages them. Make sure the planting hole is deep enough and wide enough to hold the root ball of the plant. If the plant's roots are pot-bound, carefully release them and put them in the hole at the same depth as they did in the container. To remove any air pockets, gently compact the earth surrounding the roots as you backfill the hole. After planting, give the plant a good watering to help settle the soil and supply some moisture.

Mulching is a crucial part of caring for perennial plants. A covering of organic mulch, such as wood chips, straw, or shreds of bark, helps control temperature, inhibit weed growth, and retain soil moisture. Mulch further improves soil health by adding organic matter to the soil during its decomposition. It is crucial to put mulch a few inches away from plant stems in order to avoid moisture buildup and possible rot. For most perennials, a 2 to 3-inch mulch cover is usually sufficient.

Watering is an essential part of caring for perennial plants, particularly in their establishment stage. Perennials that have just been planted need steady hydration to form robust root systems. It is more efficient to water the soil deeply—roughly 6 to 8 inches—than to water it shallowly more frequently. Many perennials become more drought-tolerant once they're established and need less frequent watering. To avoid stress and preserve plant health, more watering may be required during extended dry spells or intense heat.

Perennial plant growth and flowering are aided by fertilization. A yearly spring application of a balanced, slow-releasing fertilizer is beneficial for most perennial plants. This ensures that vital minerals like potassium, phosphorus, and nitrogen are available consistently throughout the growing season. Compost and well-rotted manure are good examples of organic fertilizers since they gradually increase the fertility and structure of the soil. In order to prevent overfertilization, which can cause excessive foliage growth at the expense of blossoms, it's critical to adhere to approved treatment rates.

Pruning is an essential part of perennial care that can help with disease prevention, healthy development, and aesthetic enhancement. Depending on the kind of perennial, different pruning schedules and techniques apply. For instance, it is recommended to prune many spring-blooming perennials—like peonies and irises—after they have finished flowering in order to straighten up the plant and remove any wasted blooms. Early spring pruning is beneficial for late-blooming perennials, such as asters and chrysanthemums, as it helps to eliminate dead stems and promote new growth. Deadheading, or pulling off faded flowers, promotes more flowering and inhibits the formation of seeds, which can take energy away from the plant.

Perennials can be effectively divided to encourage the growth of new plants, minimize overcrowding, and revitalize older ones. Every few years, the division helps many perennials retain their vigor and encourage healthy growth. When the weather is cool, and the plants are not actively growing, early spring or late autumn are the ideal seasons to divide perennials. Plants that are perennials should be carefully dug out and divided into smaller portions, each with its own roots and branches. To aid in the establishment of the new plants, replant the divisions at the same depth as before and give them plenty of water.

Managing pests and diseases is essential to keeping perennial plants healthy and productive. Aphids, slugs, snails, and spider mites are common pests, and powdery mildew, rust, and root rot are typical illnesses. Biological, chemical, and cultural controls are combined in integrated pest management (IPM) techniques, which effectively manage diseases and pests. Cultural norms that promote adequate air circulation, regular cleanliness, and proper spacing lessen the pressure from pests and diseases. Naturally controlling pest populations can be aided by biological controls, such as the introduction of beneficial insects or the use of microbial therapies. Use chemical controls, such as synthetic and organic insecticides, only as a last option and in accordance with label directions.

A sustainable and biodiverse garden can benefit greatly from the many ecological advantages provided by perennial plants. They give pollinators—like bees, butterflies, and hummingbirds—a place to live and food to eat. Pollinators are necessary for the pollination of many plants, including food crops. Additionally, beneficial insects that help reduce pest populations—like ladybugs and predatory beetles—are supported by perennials. Furthermore, because of their deep and wide root systems, perennial plants contribute significantly to soil health by lowering erosion, promoting nutrient cycling, and improving soil structure.

Perennial plants have great aesthetic benefits; they enhance gardens and landscapes with beauty, texture, and color. There are countless design options available with perennials because they may be found in a broad variety of shapes and sizes, from low-growing groundcovers to tall, erect species. Their varied hues of foliage and flowers can be combined to provide breathtaking visual harmonies and contrasts. For instance, the vivid blossoms of coneflowers can contrast with the airy plumes of ornamental grasses, while the strong leaves of hostas might be combined with the delicate flowers of astilbes. Because different species of perennials bloom at different periods during the growing season, giving a continual display of color and texture, perennials also add to seasonal interest.

Plant combinations, bloom periods, and garden structure must all be carefully planned and taken into account when designing a perennial garden. A balanced and harmonious design can be accomplished by drawing up a garden plan that specifies where perennials should be placed according to their height, color, and blooming season. Plants with comparable cultural requirements should be grouped together to guarantee proper care and growth. It's crucial to take into account the garden's general layout, combining features like borders, pathways, and focal points to make the area cohesive and visually pleasing.

One great method to maintain local ecosystems and encourage biodiversity is to plant native perennials. Because they are more suited to the local temperature and soil, native plants are less reliant on external resources like fertilizer and water. They give natural wildlife, including pollinators and helpful insects, vital habitat and food sources. Gardeners may help save native plant species and build a more ecologically balanced and sustainable landscape by using native perennials in their landscapes.

Additionally, perennial gardens can be planned with certain themes or goals in mind, including drawing pollinators, developing a cutting garden, or adding interest all year round. For instance, pollinator gardens emphasize the planting of a range of perennials that are rich in nectar and bloom continuously throughout the growing season, providing pollinators with a steady supply of food. The goal of cutting gardens is to produce an abundant supply of long-lasting and prolific bloomers for use in arrangements indoors. To ensure that a garden is beautiful throughout the year, year-round interest gardens combine a variety of perennials with varying bloom seasons, foliage colors, and textures.

Perennial gardening done wisely is a great method to reduce water use and build a landscape that is sustainable. Selecting drought-tolerant perennials, including yarrow, sedum, and lavender, helps preserve water resources by lowering the need for further irrigation. By using effective irrigation techniques, like soaker hoses or drip irrigation, water is delivered directly to the root zone, reducing runoff and evaporation. Reducing the frequency of watering and increasing water retention are achieved through mulching and adding organic matter to the soil. Plants with comparable water requirements are best grouped together to ensure they get the right amount of water without wasting any.

An excellent addition to the garden, edible perennials offer a consistent supply of food year after year. Fresh, homegrown produce can be obtained with little effort from perennial vegetables like asparagus, rhubarb, and artichokes, as well as from fruit-bearing perennials like blueberries, raspberries, and strawberries. Including edible perennials in the garden not only improves the landscape's diversity and productivity but also increases food security. These plants work well in both food and ornamental gardens since they frequently feature eye-catching leaves and blossoms.

To sum up, growing and maintaining perennial plants is a fulfilling and environmentally friendly gardening activity that has many advantages. Every stage is essential to creating a successful perennial garden, from choosing the best plants for your garden's growth conditions to putting appropriate planting and management practices into place. In addition to offering enduring beauty, perennials help local ecosystems and create a landscape that is both biodiverse and sustainable. Gardeners may develop lively, long-lasting gardens that provide aesthetic pleasure and ecological value for many years to come by learning about the requirements of perennial plants and incorporating them into well-considered garden plans.

Harvesting and preserving fruits and nuts

An age-old tradition that blends culinary arts and agricultural knowledge is harvesting and storing fruits and nuts, guaranteeing that the abundance of the growing season can be enjoyed all year round. This section explores several facets of gathering and storing fruits and nuts, such as best practices for gathering, conventional and contemporary methods of preservation, nutritional advantages, and the cultural importance of these activities. By being aware of these factors, people may maximize the yield from their gardens and orchards, cut down on waste, and enjoy wholesome foods all year round.

It is essential to harvest fruits and nuts at the optimal time of year to ensure optimal flavor, nutritional content, and storage longevity. When a fruit or nut is ready to be plucked, it will show particular signs. Apples and pears, for instance, should be picked when they are firm and have begun to take on their distinct color. To determine if an apple is ripe, gently twist it off the tree; if it comes off easily, it is ready to be picked. But in order to prevent overripening and turning mealy, pears are best harvested when they are still slightly hard and let ripen off the tree. When stone fruits, such as peaches, plums, and cherries, release slightly under light pressure and release their aroma when fully ripe, it's time to harvest them.

Strawberries, raspberries, and blueberries are examples of berries that are ready to be harvested when they are completely colored and can be pulled off the plant with ease. In order to collect berries when they are at their best and to keep overripe fruit from drawing pests, it is crucial to pick them often throughout the harvest season. Harvesting grapes while they are juicy, plump, and tasty is recommended. The ideal time to harvest the entire cluster can be ascertained by tasting a few grapes from each cluster.

Nuts have distinct harvesting indicators of their own, such as walnuts, almonds, and pecans. When the hulls separate and the nuts fall to the ground, walnuts are usually ready to be harvested. To keep fallen walnuts from spoiling, it's critical to collect them right away. When the hulls crack apart on the tree, exposing the hard shell within, it is when almonds are picked. When the shucks split open, and the nuts fall from the tree, pecans are ready to be harvested. After harvesting, nuts must be completely dried to lower moisture content and stop mold from growing while being stored.

Effective fruit and nut preservation is crucial for prolonging their shelf life and savoring their flavors long beyond the harvest season. The two main categories of preservation methods are traditional and modern, each having certain benefits and uses.

Curing, fermentation, and drying are examples of traditional preservation techniques. One of the earliest and most basic methods of preserving fruits and nuts is drying them. It entails taking the moisture out of the produce to stop spoiling microbes from growing. Fruits can be dried using a dehydrator, air drying, or sun drying techniques. You can slice and dry fruit such as apples, pears, peaches, and plums to make wholesome snacks. Berries can also be dried, albeit because of their increased moisture content, it might take them longer to dry. Usually, nuts are dried in their shells to prevent contamination and harm to the kernels. To preserve their quality, fruits and nuts that have been dried should be kept in cool, dark places in airtight containers.

Another age-old preservation technique is fermentation, which uses helpful microbes to change sugars into acids that serve as organic preservatives. Fermented fruits have a distinct sour flavor and are high in bacteria, which support digestive health. Examples of these products are sauerkraut, made from cabbage, and kimchi, made from a variety of vegetables. Fruits like pears and apples that have undergone fermentation can yield tasty drinks like perry and cider. Crushing the fruit, letting the natural or introduced yeast turn the sugars into alcohol, and finally letting the drink age to develop its flavor are all parts of the fermentation process.

Curing is a popular method of preserving nuts, especially those high in oil content, like pecans and walnuts. The process of curing entails gradually drying the nuts so as to preserve their flavor and texture while lowering their moisture level. To stop mold growth, this procedure needs to be closely watched and can take several weeks. To improve their flavor and prolong their shelf life, nuts might be roasted, salted, or flavored after they have been cured.

When it comes to preservation, modern techniques like vacuum sealing, freezing, and canning are more convenient and have longer shelf lives than older ones. Fruits and nuts are processed at high temperatures in sealed jars to eliminate enzymes and microbes that cause spoiling during the canning process. Canning can be divided into two categories: pressure canning and water bath canning. High-acid fruits such as tomatoes, berries, and stone fruits can be preserved using water bath canning; low-acid fruits and nuts require pressure canning in order to preserve freshness and avoid botulism.

A simple and adaptable way to preserve a large range of fruits and nuts is by freezing them. It entails bringing the product down to a temperature below freezing, which limits the growth of bacteria that cause spoiling and decelerates enzyme activity. To avoid clumping, fruits like berries, peaches, and cherries can be frozen separately on a baking sheet and then transferred to airtight receptacles or freezer bags. To preserve their flavor and increase their shelf life, nuts can also be frozen. Certain fruits must be blanched in order to maintain their color, texture, and nutritional content before freezing. In order to stop the cooking process, the fruit is blanched by short boiling it and then submerging it in freezing water.

Another contemporary preservation method is vacuum sealing, which is depressurizing the container to produce a vacuum-sealed atmosphere. By avoiding oxidation and lowering the possibility of spoiling, this technique greatly increases the shelf life of fruits and nuts. Nuts and fresh or blanched fruits can be stored along with dried fruits in vacuum-sealed bags or containers. When it comes to high-fat nuts that are prone to rancidity, such as walnuts and almonds, this procedure works very well for maintaining their quality.

Nuts and fruits that are preserved maintain their nutritious content in addition to having a longer shelf life. Fruits are a great source of vitamins, minerals, antioxidants, and dietary fiber—all of which are critical for overall health maintenance. Berries, for example, are a great source of phytonutrients, vitamin C, and folate, all of which enhance immune system function and reduce inflammation. Nuts are a great source of protein, healthy fats, vitamins, and minerals like selenium, magnesium, and vitamin E. These nutrients are essential for maintaining brain and heart health as well as general well-being.

Harvesting and storing fruits and nuts has profound cultural roots dating back to human history. To maximize the use of their local resources, various nations have created distinctive culinary traditions and preservation methods over the millennia. For instance, fruit and nut preservation techniques used by Native American cultures included burning and drying. In order to guarantee food supply during lean seasons, techniques like canning and fermenting gained popularity in Europe during times of abundance. In Asia, a variety of fruits and vegetables were frequently preserved using methods including pickling and salting.

Many cultures have harvest festivals and customs honoring the abundance of the harvest season. Gatherings like these frequently involve group tasks like gathering, preserving, and distributing nuts and fruits. These customs promote a feeling of belonging, appreciation, and kinship with the land. Reviving old preservation techniques and sustainable practices is becoming more popular in today's culture as individuals look to minimize food waste and reestablish a connection with their food sources.

The preservation of fruits and nuts has an impact on the environment in addition to cultural value. People can lessen their dependency on commercially processed goods, which frequently need a lot of energy and resources for production, packaging, and shipping, by canning seasonal produce. Food options produced using home preservation techniques are generally healthier and more ecologically friendly because they require fewer additives and preservatives. Preserving fruits and nuts that are grown nearby also helps local agriculture and lessens the carbon footprint that comes with shipping products across large distances.

Fruit and nut preservation is now much more effective and convenient, thanks to modern technologies. Precision and consistent harvest preservation is made easier for home gardeners and food enthusiasts by state-of-the-art dehydrators, freeze dryers, and vacuum sealers. Higher-quality preserved goods can be produced with greater control over temperature, humidity, and air exposure thanks to these instruments. In addition, anyone who is interested in learning and honing preservation techniques can find a wealth of knowledge and assistance via Internet resources, courses, and communities.

Effective fruit and nut preservation has also been aided by innovations in storage and packaging. Biodegradable or compostable bags and containers are examples of eco-friendly packaging materials that provide sustainable substitutes for conventional plastic packaging. Preserved products must be stored under ideal conditions, which include controlled temperature and protection from light and humidity. Nuts that are vacuum-sealed can have their shelf life greatly extended, and spoiling is avoided by, for instance, keeping them in the freezer and storing dried fruits in a dark, cold pantry.

Nuts and preserved fruits have a wide range of culinary uses. Dried fruits can be used as quick snacks or in baking and cooking. Examples of such fruits are figs, raisins, and apricots. To foods like oatmeal, salads, and trail mixes, they naturally impart sweetness and texture. Fermented fruits, like kimchi and sauerkraut, have a sour flavor and are high in probiotics. They are frequently used as side dishes or sauces. You can eat canned fruits on their own, as pies and cobblers, or as a way to add some sweetness to savory recipes.

When it comes to using nuts in the kitchen, they can be used raw, roasted, or in a variety of forms, such as nut kinds of butter and flour. They enhance the texture and flavor of many different foods, including baked goods, salads, sauces, and spreads. Nuts that have been preserved, like spiced pecans or almonds roasted with honey, are delicious treats and presents. Preserved fruits and nuts add taste and texture to regular meals and increase nutritional value by supplying important nutrients and healthy fats.

The skill of gathering and storing fruits and nuts is evidence of human creativity and flexibility. People may maximize the yield from their gardens and orchards, enjoy wholesome foods all year round, and support a resilient and sustainable food system by learning and practicing these techniques. The ability to preserve the harvest's abundance, whether by conventional means like drying and fermenting or by contemporary ones like freezing and vacuum sealing, is a vital one that links us to our food, our culture, and our surroundings.

CHAPTER V

Backyard Livestock

Benefits and considerations for raising animals

For thousands of years, the foundation of human civilization has been the raising of animals, which offer many advantages such as food, labor, and companionship. Animal husbandry evolved together with societies, moving from primitive means of subsistence to sophisticated agricultural techniques. This section examines the many advantages and factors to be taken into account when rearing animals, including practical, ethical, ecological, and emotional issues, along with dietary, financial, and emotional components. Comprehending these variables can aid individuals and communities in making knowledgeable choices regarding animal husbandry while also advocating for humane and sustainable methods.

Animal products, including meat, milk, eggs, and fish, provide vital nutrients, which is one of the main advantages of animal husbandry. These goods are abundant in essential vitamins, minerals, and high-quality proteins that are vital to human health. For instance, beef supplies iron, vitamin B12, and key amino acids, all of which are necessary for brain function, blood oxygen delivery, and muscular growth. The minerals, calcium, phosphorus, and vitamin D found in dairy products—such as milk, cheese, and yogurt—support healthy bones and metabolic functions. Eggs are a nutrient-dense, adaptable food that is high in protein, choline, and other vitamins and minerals. Omega-3 fatty acids, which are advantageous for cardiovascular health and cognitive function, are abundant in fish and seafood.

Beyond the nutritional advantages, keeping animals has a major positive impact on economic growth and stability, especially in rural and agricultural areas. Millions of people around the world, from small-scale farmers to major commercial operations, rely on livestock farming for their livelihood. The money received from the sale of animal products goes toward raising living standards, investing in education, and supporting families. Furthermore, animals are essential to integrated agricultural systems because they supply labor for field preparation, manure for crop fertilization, and goods for trade or sale. Livestock and crop production have a synergistic relationship that increases farm productivity and sustainability.

If done properly, raising animals has many positive ecological effects as well. Through grazing, animals help to maintain grasslands, pastures, and other ecosystems by preventing vegetation from growing out of control, lowering the risk of wildfires, and fostering biodiversity. Cattle, sheep, and goats are examples of grazing livestock that aid in the management of invasive plant species and promote the establishment of native grasses and forbs. By encouraging the growth of robust root systems and boosting nutrient cycling, this grazing practice can also enhance soil health. Animal dung is an important organic fertilizer in mixed agricultural systems since it improves the soil's structure and nutrient content. The use of synthetic fertilizers, which may have detrimental effects on the environment, is lessened by this natural fertilization.

Moreover, when combined with sustainable land management techniques, animal husbandry can help sequester carbon and slow down global warming. For example, silvopastoral systems—which integrate cattle with trees and shrubs—and well-managed grazing systems can improve the amount of carbon stored in soils and vegetation. These methods support overall climate resilience by mitigating greenhouse gas emissions from cattle.

Numerous non-food goods that are necessary for a variety of industries and daily living are also provided by animals. Textiles, footwear, and apparel are made from wool, leather, and other animal fibers. Products like gelatin, adhesives, and fertilizers are made from animal byproducts, including bones, hooves, and hides. Animals are also used in biological research and in the creation of medications, vaccinations, and other healthcare items. The adaptability and usefulness of goods obtained from animals highlight how crucial animal husbandry is to sustaining a variety of economic sectors and addressing human needs.

Animals provide enormous emotional and psychological benefits in addition to their practical and financial advantages. Dogs, cats, and horses are examples of companion animals that offer company, lower stress levels, and enhance mental health and well-being. The relationship between people and animals can strengthen interpersonal relationships, lessen feelings of isolation, and encourage a sense of accountability and purpose. Service animals and therapy animals are educated to help people with mental, emotional, and physical impairments, enhancing their freedom and quality of life. It has been demonstrated that having animals in therapeutic settings—like hospitals, nursing homes, and rehabilitation centers—reduces anxiety, lessens sadness, and improves general well-being.

Raising animals can be a very useful educational tool for imparting life skills, empathy, and responsibility. Programs such as Future Farmers of America (FFA) and 4-H provide young people with practical skills in animal husbandry and care in educational settings. These courses impart useful knowledge about biology, agriculture, and environmental stewardship while also encouraging a feeling of accountability, leadership, and teamwork. People can get an understanding of the intricacies of life cycles, ecosystems, and the interconnectedness of living things by providing care for animals.

Raising animals has many advantages, but in order to maintain sustainability and ethical treatment of animals, there are significant issues and difficulties that need to be taken into account. Animal welfare, which includes the mental and physical health of animals, is one of the main issues. The provision of sufficient food, water, housing, and veterinary care, together with the reduction of stress and suffering, are the top priorities in ethical animal husbandry. It is crucial to design habitats that let animals engage in activities that come naturally to them, like grazing, socializing, and foraging. Freedom from hunger and thirst, freedom from discomfort, freedom from pain, damage, and sickness, freedom to exhibit normal behavior, and freedom from fear and distress are all included in the Five Freedoms framework, which lays forth essential guidelines for animal welfare.

Because of its frequently subpar circumstances and effects on animal welfare, intensive animal farming, also known as factory farming, has given rise to serious ethical and environmental problems. In intensive farming systems, activities including overcrowding, confinement, and the use of growth hormones and antibiotics are frequent and can be harmful to the health and welfare of the animals. Furthermore, intensive farming's negative environmental effects, such as deforestation, greenhouse gas emissions, and water pollution, provide serious obstacles to sustainability. Moving toward more sustainable and compassionate farming methods, like organic farming, regenerative agriculture, and pasture-based systems, is necessary to address these problems.

Raising animals also requires careful attention to the environment, especially in light of resource management and climate change. Methane emissions from enteric fermentation in ruminants (cattle, sheep, and goats) and nitrous oxide emissions from waste management are the main ways that livestock husbandry contributes to greenhouse gas emissions. Enhancing feed efficiency, enhancing grazing techniques, and utilizing technology like anaerobic digesters to extract and use methane from manure are among the strategies to reduce these emissions. Water management and utilization are also important concerns because animal husbandry uses a lot of water for washing, drinking, and producing feed. Reducing the environmental impact of animal husbandry can be achieved by putting effective water management systems and conservation techniques into operation.

When it comes to animal production, biodiversity and land use are also crucial factors. Ecosystem imbalances, biodiversity loss, and soil degradation can result from overgrazing and habitat damage. Ecologically sound grazing methods, like rotating grazing and incorporating a variety of plant species, can promote biodiversity and preserve ecosystems. Preserving native species and fostering ecological resilience requires the protection of natural ecosystems and the implementation of conservation measures like wildlife corridors and buffer zones.

The practical aspects of animal husbandry involve the identification of suitable species and breeds, comprehension of their individual requirements, and provision of adequate infrastructure and resources. The needs for space, food, climate, and care vary throughout species and breeds. For instance, housing and management techniques for growing chickens for egg production differ from those for producing cattle for beef. Selecting species and breeds that are compatible with the local climate and available resources is crucial. Animal health and safety depend on having the right infrastructure, which includes feeding programs, fencing, and housing. To preserve animal health and stop the transmission of infectious diseases, routine veterinarian care and disease prevention techniques, including as immunizations and biosecurity procedures, are essential.

The husbandry of animals also heavily involves economic factors. To guarantee profitability and sustainability, the expenses associated with animal husbandry, such as feed, veterinary care, equipment, and labor, need to be properly controlled. The financial sustainability of animal farming businesses can be impacted by market demand, pricing, and competition. Enhancing economic resilience can be achieved through diversifying revenue streams, such as by selling value-added products (such as cheese and wool products) or providing agritourism experiences. The ability of animal farming businesses to access markets, infrastructure, and support services, including cooperative networks and extension programs, is crucial to their success.

In animal husbandry, legal and regulatory factors are equally crucial. Ethical and sustainable farming operations depend on adherence to environmental rules, food safety requirements, and laws pertaining to animal care. Complying with these standards and keeping an awareness of them will assist in avoiding legal problems and improve the marketability and reputation of animal products. Participating in certification schemes, such as organic or humane certification, can also reassure customers regarding the caliber and moral principles of animal products.

There are important cultural and societal ramifications to animal rearing as well. Animals have always been important to many different cultural customs, rites, and social institutions. Animals are frequently included in mythology, artwork, and ceremonial practices in numerous indigenous cultures, where they are seen as essential components of the natural world. The way that people and animals interact reflects broader cultural values and beliefs about community, sustainability, and the natural world. Due to growing awareness of animal welfare, environmental sustainability, and food security, ethical and sustainable animal husbandry practices are becoming more and more important in today's society.

In summary, there are several advantages to keeping animals, including improved nutrition, financial stability, and psychological and ecological health. To ensure moral, sustainable, and responsible actions, it also brings up important issues and difficulties that need to be resolved. Individuals and communities can make knowledgeable decisions regarding animal husbandry by putting an emphasis on animal welfare, using sustainable farming practices, and comprehending the intricate interactions between environmental, economic, and social variables. The holistic method of animal husbandry not only improves the lives of humans and animals alike, but it also strengthens and sustains the food chain. The practice of animal husbandry can change to meet the demands of a changing world while respecting the close relationships that exist between people, animals, and the environment through ongoing learning, innovation, and cooperation.

Chickens: breeds, care, and egg production

For thousands of years, the foundation of human civilization has been the raising of animals, which offer many advantages such as food, labor, and companionship. Animal husbandry evolved together with societies, moving from primitive means of subsistence to sophisticated agricultural techniques. This section examines the many advantages and factors to be taken into account when rearing animals, including practical, ethical, ecological, and emotional issues, along with dietary, financial, and emotional components. Comprehending these variables can aid individuals and communities in making knowledgeable choices regarding animal husbandry while also advocating for humane and sustainable methods.

Animal products, including meat, milk, eggs, and fish, provide vital nutrients, which is one of the main advantages of animal husbandry. These goods are abundant in essential vitamins, minerals, and high-quality proteins that are vital to human health. For instance, beef supplies iron, vitamin B12, and key amino acids, all of which are necessary for brain function, blood oxygen delivery, and muscular growth. The minerals, calcium, phosphorus, and vitamin D found in dairy products—such as milk, cheese, and yogurt—support healthy bones and metabolic functions. Eggs are a nutrient-dense, adaptable food that is high in protein, choline, and other vitamins and minerals. Omega-3 fatty acids, which are advantageous for cardiovascular health and cognitive function, are abundant in fish and seafood.

Beyond the nutritional advantages, keeping animals has a major positive impact on economic growth and stability, especially in rural and agricultural areas. Millions of people around the world, from small-scale farmers to major commercial operations, rely on livestock farming for their livelihood. The money received from the sale of animal products goes toward raising living standards, investing in education, and supporting families. Furthermore, animals are essential to integrated agricultural systems because they supply labor for field preparation, manure for crop fertilization, and goods for trade or sale. Livestock and crop production have a synergistic relationship that increases farm productivity and sustainability.

If done properly, raising animals has many positive ecological effects as well. Through grazing, animals help to maintain grasslands, pastures, and other ecosystems by preventing vegetation from growing out of control, lowering the risk of wildfires, and fostering biodiversity. Cattle, sheep, and goats are examples of grazing livestock that aid in the management of invasive plant species and promote the establishment of native grasses and forbs.

By encouraging the growth of robust root systems and boosting nutrient cycling, this grazing practice can also enhance soil health. Animal dung is an important organic fertilizer in mixed agricultural systems since it improves the soil's structure and nutrient content. The use of synthetic fertilizers, which may have detrimental effects on the environment, is lessened by this natural fertilization.

Moreover, when combined with sustainable land management techniques, animal husbandry can help sequester carbon and slow down global warming. For example, silvopastoral systems—which integrate cattle with trees and shrubs—and well-managed grazing systems can improve the amount of carbon stored in soils and vegetation. These methods support overall climate resilience by mitigating greenhouse gas emissions from cattle.

Numerous non-food goods that are necessary for a variety of industries and daily living are also provided by animals. Textiles, footwear, and apparel are made from wool, leather, and other animal fibers. Products like gelatin, adhesives, and fertilizers are made from animal byproducts, including bones, hooves, and hides. Animals are also used in biological research and in the creation of medications, vaccinations, and other healthcare items. The adaptability and usefulness of goods obtained from animals highlight how crucial animal husbandry is to sustaining a variety of economic sectors and addressing human needs.

Animals provide enormous emotional and psychological benefits in addition to their practical and financial advantages. Dogs, cats, and horses are examples of companion animals that offer company, lower stress levels, and enhance mental health and well-being. The relationship between people and animals can strengthen interpersonal relationships, lessen feelings of isolation, and encourage a sense of accountability and purpose. Service animals and therapy animals are educated to help people with mental, emotional, and physical impairments, enhancing their freedom and quality of life. It has been demonstrated that having animals in therapeutic settings—like hospitals, nursing homes, and rehabilitation centers—reduces anxiety, lessens sadness, and improves general well-being.

Raising animals can be a very useful educational tool for imparting life skills, empathy, and responsibility. Programs such as Future Farmers of America (FFA) and 4-H provide young people with practical skills in animal husbandry and care in educational settings. These courses impart useful knowledge about biology, agriculture, and environmental stewardship while also encouraging a feeling of accountability, leadership, and teamwork. People can get an understanding of the intricacies of life cycles, ecosystems, and the interconnectedness of living things by providing care for animals.

Raising animals has many advantages, but in order to maintain sustainability and ethical treatment of animals, there are significant issues and difficulties that need to be taken into account. Animal welfare, which includes the mental and physical health of animals, is one of the main issues. The provision of sufficient food, water, housing, and veterinary care, together with the reduction of stress and suffering, are the top priorities in ethical animal husbandry. It is crucial to design habitats that let animals engage in activities that come naturally to them, like grazing, socializing, and foraging. Freedom from hunger and thirst, freedom from discomfort, freedom from pain, damage, and sickness, freedom to exhibit normal behavior, and freedom from fear and distress are all included in the Five Freedoms framework, which lays forth essential guidelines for animal welfare.

Because of its frequently subpar circumstances and effects on animal welfare, intensive animal farming, also known as factory farming, has given rise to serious ethical and environmental problems. In intensive farming systems, activities including overcrowding, confinement, and the use of growth hormones and antibiotics are frequent and can be harmful to the health and welfare of the animals. Furthermore, intensive farming's negative environmental effects, such as deforestation, greenhouse gas emissions, and water pollution, provide serious obstacles to sustainability. Moving toward more sustainable and compassionate farming methods, like organic farming, regenerative agriculture, and pasture-based systems, is necessary to address these problems.

Raising animals also requires careful attention to the environment, especially in light of resource management and climate change. Methane emissions from enteric fermentation in ruminants (cattle, sheep, and goats) and nitrous oxide emissions from waste management are the main ways that livestock husbandry contributes to greenhouse gas emissions. Enhancing feed efficiency, enhancing grazing techniques, and utilizing technology like anaerobic digesters to extract and use methane from manure are among the strategies to reduce these emissions. Water management and utilization are also important concerns because animal husbandry uses a lot of water for washing, drinking, and producing feed. Reducing the environmental impact of animal husbandry can be achieved by putting effective water management systems and conservation techniques into operation.

When it comes to animal production, biodiversity and land use are also crucial factors. Ecosystem imbalances, biodiversity loss, and soil degradation can result from overgrazing and habitat damage. Ecologically sound grazing methods, like rotating grazing and incorporating a variety of plant species, can promote biodiversity and preserve ecosystems. Preserving native species and fostering ecological resilience requires the protection of natural ecosystems and the implementation of conservation measures like wildlife corridors and buffer zones.

The practical aspects of animal husbandry involve the identification of suitable species and breeds, comprehension of their individual requirements, and provision of adequate infrastructure and resources. The needs for space, food, climate, and care vary throughout species and breeds. For instance, housing and management techniques for growing chickens for egg production differ from those for producing cattle for beef. Selecting species and breeds that are compatible with the local climate and available resources is crucial. Animal health and safety depend on having the right infrastructure, which includes feeding programs, fencing, and housing. To preserve animal health and stop the transmission of infectious diseases, routine veterinarian care and disease prevention techniques, including as immunizations and biosecurity procedures, are essential.

The husbandry of animals also heavily involves economic factors. To guarantee profitability and sustainability, the expenses associated with animal husbandry, such as feed, veterinary care, equipment, and labor, need to be properly controlled. The financial sustainability of animal farming businesses can be impacted by market demand, pricing, and competition. Enhancing economic resilience can be achieved through diversifying revenue streams, such as by selling value-added products (such as cheese and wool products) or providing agritourism experiences. The ability of animal farming businesses to access markets, infrastructure, and support services, including cooperative networks and extension programs, is crucial to their success.

In animal husbandry, legal and regulatory factors are equally crucial. Ethical and sustainable farming operations depend on adherence to environmental rules, food safety requirements, and laws pertaining to animal care. Complying with these standards and keeping an awareness of them will assist in avoiding legal problems and improve the marketability and reputation of animal products. Participating in certification schemes, such as organic or humane certification, can also reassure customers regarding the caliber and moral principles of animal products.

There are important cultural and societal ramifications to animal rearing as well. Animals have always been important to many different cultural customs, rites, and social institutions. Animals are frequently included in mythology, artwork, and ceremonial practices in numerous indigenous cultures, where they are seen as essential components of the natural world. The way that people and animals interact reflects broader cultural values and beliefs about community, sustainability, and the natural world. Due to growing awareness of animal welfare, environmental sustainability, and food security, ethical and sustainable animal husbandry practices are becoming more and more important in today's society.

In summary, there are several advantages to keeping animals, including improved nutrition, financial stability, and psychological and ecological health. To ensure moral, sustainable, and responsible actions, it also brings up important issues and difficulties that need to be resolved. Individuals and communities can make knowledgeable decisions regarding animal husbandry by putting an emphasis on animal welfare, using sustainable farming practices, and comprehending the intricate interactions between environmental, economic, and social variables. The holistic method of animal husbandry not only improves the lives of humans and animals alike, but it also strengthens and sustains the food chain. The practice of animal husbandry can change to meet the demands of a changing world while respecting the close relationships that exist between people, animals, and the environment through ongoing learning, innovation, and cooperation.

Bees: starting a beehive and honey production

Beekeeping and honey production have long been prized for their sweet benefits, as well as the vital role that bees perform in pollinating crops and maintaining biodiversity. This section delves into the complex world of beekeeping, covering everything from the basic procedures for establishing a hive to the complexities involved in producing honey and its wider effects on the environment and agriculture. Comprehending these facets is crucial for individuals who aspire to become beekeepers or are engaged in sustainable food production and ecological well-being.

It takes meticulous planning and preparation to start a beehive. Selecting a good site for the hive is the first step. The hive should ideally be situated in a sunny area that offers some shade during the warmest hours of the day. It should also have access to a nearby water source and be protected from severe winds. In order to give their bees plenty of food, beekeepers frequently erect hives in orchards, gardens, or next to fields of wildflowers. Since pesticides can damage bees and taint honey, it is important to make sure that no neighboring crops have been treated with pesticides.

Next, it's crucial to choose the right kind of hive and accessories. Because of its uniform frame sizes and simplicity of maintenance, the Langstroth hive is the most popular type of hive used in beekeeping. Some hive designs, like Warre and top-bar hives, provide alternative management approaches and can be of interest to beekeepers who prefer natural or unconventional techniques. In addition to the hive, beekeepers require basic supplies such as feeders for extra feeding as needed, a smoker, hive tools, and protective clothes like a beekeeping suit or jacket, veil, and gloves.

Buying a package of bees or a nucleus colony (nuc) from a reliable bee breeder or supplier is the usual method of getting bees for the hive. A nuc consists of a queen bee, worker bees, and frames of brood (developing bees) and honey, whereas a package of bees has a queen bee and several thousand worker bees. When bees are installed into a hive, they must be cautiously brought to their new surroundings, with special attention paid to releasing the queen securely and providing food and water for the swarm. To reduce stress and guarantee that the bees establish themselves in the hive, this procedure calls for patience and cautious treatment.

After the hive is created, it requires constant attention and management to keep bee colonies healthy and promote honey production. The colony's health, the evaluation of honey storage, and the detection of pest or disease indicators all depend on routine hive inspections. Beekeepers look for the queen during inspections, as well as the general population size of the colony and the brood-laying tendency. Additionally, they could switch around the frames inside the hive to encourage uniform brood growth and honey yield.

A crucial part of beekeeping is controlling diseases and pests to maintain the hive's long-term health and productivity. Wax moths, tiny hive beetles, and varroa mites are common pests that harm honey bees. Varroa mites are especially harmful because they feed on bee colonies' hemolymph, or blood, and spread viruses to them. Employing screened bottom boards, employing organic treatments, and adopting brood control techniques are a few examples of integrated pest management (IPM) practices that help minimize chemical interventions while mitigating pest problems.

If left untreated, diseases like European and American foulbrood (EFB) can completely destroy bee colonies. In order to stop the development of these diseases, beekeepers should be aware of their symptoms and follow stringent hygiene precautions, such as sterilizing equipment between hives and maintaining proper apiary hygiene. In order to prevent outbreaks and save other bee populations, certain areas may impose quarantines or the destruction of hives on beekeepers who fail to report suspected cases of foulbrood.

When nectar is scarce, or the weather is bad, feeding bees makes sure they have enough food to survive and continue raising their young. Made by combining water and granulated sugar, sugar syrup gives bees the vital carbohydrates they need for energy and is frequently used as supplemental feeding. When natural pollen supplies are scarce, patties or pollen substitutes can also be fed to encourage brood development and colony growth. Beekeepers are better able to judge when feeding is required and modify feeding procedures when hive weight and honey storage are monitored.

Harvesting and preparing honey from the hive is the satisfying part of beekeeping known as honey production. The strength of the colony, the local nectar flows, and the weather all affect when honey should be harvested. Beekeepers usually hold off until the bees have filled and sealed the extra hive boxes, or honey supers, with beeswax. This suggests that the honey is ready to be harvested and has a low moisture level.

The first step in the honey harvesting procedure is taking the honey supers out of the hive and moving them to a processing center. Before uncapping the honeycomb cells, beekeepers use a bee brush or blower to carefully remove any bees from the frames. The honey within can be accessed by uncapping the beeswax cappings using a heated knife or uncapping fork. After being uncapped, the frames are put into a centrifugal equipment called a honey extractor, which spins the frames to remove honey from the comb.

After being extracted, the honey is filtered to get rid of any last bits of debris and beeswax before being kept in jars or buckets that are suitable for food storage. To maintain its flavor and freshness, honey that has been properly preserved should be kept out of direct sunlight in a cool, dry location. While some beekeepers decide to utilize their honey for presents or personal consumption, others may decide to sell it locally, at farmers' markets, or online.

In addition to producing honey, bees are essential for pollinating crops and maintaining biodiversity. A lot of flowering plants, including fruits, vegetables, and nuts, need pollination to reproduce. One-third of the food crops grown worldwide are produced by honey bees, native bees, and other pollinators. Because of their contribution to agriculture, bee populations need to be preserved, and bee-friendly measures like growing pollinator-friendly plants, using fewer pesticides, and supporting habitat conservation initiatives need to be encouraged.

Environmental variables that affect bee health and population dynamics include illnesses, pesticide exposure, habitat loss, and climate change. The rapid and extensive collapse of honey bee colonies, known as Colony Collapse Disorder (CCD), has sparked worries about the long-term viability of beekeeping and the services it provides for agricultural pollination. In order to enhance bee health and resilience, beekeepers, academics, policymakers, and the general public must work together to address these concerns.

Sustainable beekeeping operations must include outreach and education. Beekeepers can stay up to date on cutting-edge techniques, new threats, and best management practices by taking advantage of continuing education options offered by beekeeping associations, workshops, and classes. Collaboration and a sense of community are fostered by exchanging information and experiences with other beekeepers, which helps all beekeeping activities succeed together.

In conclusion, beekeeping is a worthwhile and fulfilling hobby for anyone with an interest in pollinator conservation, sustainable agriculture, or honey production. It takes careful planning, the right tools, knowledge of bee behavior, and experience with management techniques to start a beehive. Beekeepers may maintain healthy bee colonies, improve the health of the ecosystem, and ensure food security by engaging in appropriate beekeeping practices, such as managing hives, controlling pests and diseases, and producing honey. Beekeepers are critical to maintaining bee numbers and their vital role in agricultural and natural ecosystems because they support bee-friendly conditions and advocate for pollinator conservation.

CONCLUSION

The book "Homesteading in Your Backyard: Harnessing Nature's Bounty Right Outside Your Door" captures the ageless charm and sensible advice of adopting homesteading techniques for sustainable living. Readers are taken on a self-sufficient trip that includes everything from planting gardens and rearing animals to gathering honey and canning fruits throughout the book. Each chapter is infused with the author's profound understanding and enthusiasm for homesteading, providing a combination of insightful inspiration and useful guidance.

Fundamentally, the book highlights the strong relationship that exists between people and the natural environment, promoting resource conservation and encouraging a greater understanding of the advantages of living nearer to the natural world. It encourages readers to adopt a way of living that places an emphasis on environmental awareness, sustainability, and fortitude in the face of contemporary difficulties.

With thorough instructions on choosing crops, taking care of livestock, and canning harvests, "Homesteading in Your Backyard" gives readers the skills and information they need to confidently start their homesteading journey. The book presents useful answers suited to various surroundings and lifestyles, whether readers are country enthusiasts extending their self-sufficiency endeavors or urban people trying to make the most of their little spaces.

Beyond helpful advice, the book highlights the pleasures of homesteading, such as the gratification that comes from producing your own food, gathering honey from backyard bees, and maintaining a healthy ecosystem just outside your door. It exhorts readers to adopt a more straightforward, environmentally friendly lifestyle that respects customs while adjusting to contemporary demands.

In summary, anyone interested in homesteading, gardening, or sustainable living will find "Homesteading in Your Backyard: Harnessing Nature's Bounty Right Outside Your Door" to be an extensive resource and a source of inspiration. It challenges readers to rediscover their relationship with the land, develop resilience, and enjoy the benefits of utilizing nature's abundance in their own backyard.

Thank you for buying and reading/ listening to our book. If you found this book useful/ helpful please take a few minutes and leave a review on the platform where you purchased our book. Your feedback matters greatly to us.

www.ingramcontent.com/pod-product-compliance
Lightning Source LLC
Chambersburg PA
CBHW050438010526
44118CB00013B/1583